# Aussie Airmen

## From Box Kite to Boeing

# Written & illustrated by Murray McLeod

Copyright © Murray McLeod 2018

All rights reserved. This book is copyright protected. Apart from any fair dealings for the purpose of private study, criticism, research or review as permitted under the *Copyright Act (Australia)*, no part may be reproduced by any process without written permission from the copyright owner.

Original Artwork © by the Author, Murray McLeod
Photo reference: Wings magazine

This book, and others by Murray McLeod may be purchased on [www.amazon.com](www.amazon.com) online bookstores and other retailers.

## Other titles

### Children's books
Tom, the train spotter

### Nonfiction
Aces and Adventurers
Aussie Tennis Greats
Flying Matilda
For Valour
Images of Eagles
Moto Gp
Tales of the Unexpected
The Unapproachable Norton
TT Legends
The Blue Max
Air VCs of World War One

### Fiction
Elliot's Odyssey
The Pilgrimage

- William Hart…..Aviator's.Licence No.1……………………………..8
- John Duigan…First Locally-built Aircraft…………………...........11
- Frank McNamara  First Aussie Air VC…………………………......15
- Harry Cobby….Top-scoring.AFC Ace ………………………..........19
- Ross and Keith Smith.. England-Australia Flight………………....23
- Bert Hinkler……Australia's Lone Eagle………………… ………...29
- Kingsford Smith…1928 Pacific Flight………………………………34
- Keith Anderson…Kookaburra 1929………………………………...44
- Southern Cloud…1931 Loss and Discovery……………………….50
- Charles Ulm…Stella Australis 1934…………………………….....55.
- P.G Taylor………1935 Jubilee Mail Flight………. ………….....61
- Lady Southern Cross……Smithy's Last Flight…………………...67
- DH86……Unnamed and Unloved…………………………………...75
- Bernard O'Reilly…….The Missing Stinson 1937…………………..81
- Les Clisby…Hurricane Ace 1940…………………………………...87
- Hughie Edwards VC 1941………………………………………….91
- Ron Middleton …VC 1942…………………………………………..97
- Keith Truscott…Spitfire Ace………………………………………102
- Bill Newton VC 1943……………………………………………....106
- Clive Caldwell…Top-scoring Aussie Ace………………………..110
- Donald Bennett…..Master Airman…………………………….....115
- G.U. (Scotty) Allan…The Flying Scotsman…….. …………….....121

# Aussie Airmen

The beginning of the twentieth century was a time of great change for many nations. In 1901 Australia began a federation of its six states; New South Wales, Victoria, Queensland, Tasmania, South Australia, Western Australia, this creating the Commonwealth of Australia, while the Northern Territory for the moment maintained its independence. Australia remained part of the British Empire, which at the time included Canada, New Zealand, India, South Africa, plus a mix of smaller nations in Africa and islands scattered across the globe.

Travel between nations was of necessity by ship and for the interstate traveller by train and for shorter journeys by horse-drawn vehicles or on horseback. The motor car was a rare sight on the highways or towns in those early days, and these vehicles, looking quite basic to modern eyes were mostly acquired by the upper strata of society; notably the medical profession, the Law, business leaders and politicians.

In The United States, two brothers, Wilbur and Orville Wright, bicycle makers from Dayton, Ohio, had been conducting experiments with kites and man-carrying gliders. Armed with knowledge of aeronautics they directed their energies to a powered version of those gliders. After three years of hard work, scientific reasoning and sheer determination their 'Flying Machine' was deemed ready to show its potential.

December 17 1903 should have been a date etched in history, when Orville Wright coaxed their stick, wire and canvas contraption into the air at the isolated settlement of Kitty Hawk, North Carolina. By modern standards it was not much of a flight; Orville covered only 120 feet, less than the wingspan of many current airliners; the actual flight lasted just 12 seconds, but from that modest beginning have come wings to lift the world. This should have been the story of a lifetime for the American press, yet only a handful of newspapers found room for a garbled, half-sceptical report of an experiment at Kitty Hawk, North Carolina.

Such a ground-breaking event should have been of great satisfaction for the Wright brothers, but even in America their claim was disputed for an unreasonable amount of time. Eventually they were accepted, while in America and Europe other designers were quick to follow the lead created by Wilbur and Orville Wright.

Despite its isolation from America and Europe, situated as it is, 12000 miles from Mother England, Australia began an early association with this exciting new element of air travel. Our book takes the reader through Australia's involvement, from those first faltering steps in the early part of the past century to the dawn of the jet age.

## Tribute

In taking to the air those early airmen were making a virtual step into the unknown. They flew by instinct in daunting situations, with the barest of technical aids, and in aircraft with no margins for error. Without the ability to climb above bad weather and lack of radio communication it was a tribute to their perseverance that those pioneering years did not produce more disasters.

# Pilot's Licence No.1 (1911)

**W.E.(Bill) Hart**

# William Hart (1885-1943)

The prestige of being awarded Australian Aviator's Licence No.1 went to William Ewart (Bill) Hart. Born 1885 at Parramatta New South Wales, he was the third of nine children of William Hart, timber merchant. At age 16 Hart began a dental apprenticeship and on completion of the course, he went on to open surgeries in Sydney and at country centres.

Another interest was aviation, then in its infancy in Australia, however in 1911, being well-placed financially he was able to purchase a Bristol Boxkite, which was being demonstrated in Australia by Bristol's test pilot Joseph Hammond. The price of this machine, which resembled later versions of the Wright brothers 'Flyer' was £1,333, a substantial sum of money in those days and enough to purchase two modest bungalows.

After receiving some tuition from Hammond's mechanic, Hart first flew solo on 3 September 1911, and following completion of flying tests for the Aerial League of Australia he was presented with Australian Aviator's Licence No.1, dated 5 December 1911. By early 1912 Hart had completed a number of flights around New South Wales and in January he opened a flying school at Penrith, later transferring the operation to Richmond.

Keen to build his own aeroplane, Hart constructed a two-seat monoplane of his own design and in August 1912 it was successfully test flown at Wagga Wagga, however the project had a bad outcome when the machine was wrecked in a crash at Richmond in September. Bill Hart was seriously injured in the accident and never flew again.

Following the outbreak of war in August 1914 Hart enlisted in the Australian Imperial Force and in January 1916 was installed as a lieutenant in No.1 squadron Australian Flying Corps, going on to serve in Egypt and England as an instructor. He was later diagnosed as medically unfit and in September 1916 was discharged and returned to Australia.

Post war he expanded his dental practice, opening spacious new surgeries in Sydney, and during the 1930s he toured Britain and America to observe the latest dental procedures. In World War 2 Hart again volunteered for service in the Air Force but was once more rejected as medically unfit. William Hart died in 1943, aged 58, a well-remembered pioneer airman and in 1963 a memorial to him was unveiled at Parramatta Park, an appropriate venue for one of its pioneering citizens.

**Bristol Box Kite**

# First locally-built aircraft (1910)

**John Duigan**

# John Duigan (1882-1951)

Victorian engineer John Duigan gained fame as the first Australian to construct and fly a locally-built aircraft. Born at Terang Victoria, Duigan was educated at Brighton Grammar School, and on completion, he travelled to England in 1901 to study electrical engineering, and following this, he went on to complete a course in motor engineering. Returning to Australia in 1907 he went on to live at his father's sheep station at Mia Mia in central Victoria.

Inspired by the achievements of the Wright brothers, Duigan began his own aviation experiments, beginning with a glider, based on the Wright design. This was followed by something more ambitious, a powered aircraft fitted with a four-cylinder engine locally made in Melbourne. The design, similar in layout to the French Henri Farman machines made a short flight on 16 July 1910, although Duigan did not consider this to be fully controlled. On 7 October 1910 he made a successful flight, followed by longer and higher ones.

John Duigan's younger brother Reg. was an active participant in the construction and flight testing and later flew it on several occasions. On 3 May 1911 the biplane was flown in public at Bendigo racecourse, completing a circling flight above a crowd of interested spectators. Duigan returned to England in June 1911 to gain his aviator's certificate (No.211) which was achieved at Brooklands in April 1912. In late 1911 Duigan had purchased a basic tractor biplane designed by Avro for his flights, and prior to returning to Australia the Avro was sold.

At their parent's Melbourne home the brothers designed and built a lighter version of the Avro. Unfortunately on its first flight on 17 December 1913 at Keilor Plains the machine crashed, leaving Duigan badly bruised. Following the outbreak of war Duigan enlisted in the Australian Flying Corps and in March 1916 was commissioned lieutenant and assigned to No.3 squadron. This unit operated in the photo-reconnaissance role using the two-seat RE8 biplane. Promoted to captain, Duigan saw action in France from December 1917 to May 1918, and during a photo mission his lone RE8 was set on by four German DR1 triplanes. Despite being severely wounded Duigan managed to return to base, saving the life of his wounded observer Lt. Patterson. For this remarkable display of skill and bravery Duigan was awarded the Military Cross.

Surviving his wounds, Duigan was demobilised in 1919 and returned to Melbourne to resume his profession as an engineer. He established his own garage business at Yarrawonga in 1928, until returning to Melbourne in 1941.

During World War 2 he served as an inspector of aircraft parts until the war's ending, following which he retired to Melbourne. John Duigan died in 1951 and his memory was perpetuated in 1960 with a memorial near the site of his first flight, fifty years previously, also with a 6 cents postage stamp which portrays the brothers Reg. and John Duigan.

**RE8 (No.3 Sqdrn. AFC)**

# First Australian air VC (1917)

**Frank McNamara**

# Frank McNamara 1894-1961

During World War 1 nineteen British airmen were recipients of the supreme award for gallantry, the Victoria Cross and one Australian features in such exalted company. He was Frank McNamara, a pilot serving in Palestine with No.1 squadron Australian Flying Corps during 1917. Born in Rushworth Victoria in 1894, McNamara was the eldest of eight children and on graduating from teachers college in 1913 he taught at various Melbourne schools. He was also commissioned lieutenant in a local militia unit and at the declaration of war in August 1914 McNamara was mobilised for service.

After passing through Officer's Training School he served as an instructor until February 1915, when McNamara was an early volunteer for flying training at Point Cook. He made his first solo flight in September, graduating as a pilot in October 1915. In March 1916 McNamara and his draft sailed to the Middle East, arriving at Suez in April, and from there he was seconded to England to attend Central Flying School at Upavon.

With that course completed McNamara was posted back to Egypt and for a time served as a flying instructor before joining No.1 squadron AFC, which was based in Sinai on photo-reconnaissance and bombing missions. This unit operated with the obsolete Be2 two-seater and single-seat Martynside, and on 20 March 1917 McNamara was one of four No1 pilots taking part in a raid on a Turkish rail junction near Gaza.

McNamara's Martinsyde was armed with modified howitzer shells, three of which were dropped successfully, however the fourth exploded prematurely in the cockpit, severely wounding McNamara in the leg. Returning to base McNamara spotted a fellow pilot, Lt. Rutherford on the ground beside his crash-landed Be2. Turkish cavalry could be seen fast approaching the site, and despite the rough terrain and his leg wound McNamara landed near Rutherford in a hazardous rescue attempt.

With no spare cockpit in the Martynside, Rutherford was forced to grasp a wing strut while McNamara began his take-off. This all ended badly with McNamara losing control due to his leg wound, plus Rutherford's weight overbalancing the aircraft. The intrepid pair escaped further injury, and after setting fire to the Martynside they dashed to the Be2 where Rutherford repaired the engine, while McNamara opened fire on the approaching cavalry with his service revolver.

They also received assistance from two of the other pilots who began strafing the enemy troops, while McNamara started the engine and with Rutherford installed in the second cockpit they managed to make a dramatic escape.

Despite the agony of his wound and loss of blood McNamara remained conscious and flew the damaged aircraft the 70km back to base at El Arish. For his gallantry of 20 March McNamara was awarded the Victoria Cross, although this was not officially invested until 20 May 1921 by the Prince of Wales at a ceremony at Government House Melbourne.

Post war Frank McNamara enlisted as a flying officer in the newly-established Royal Australian Air Force and at the start of World War 2 held the rank of air commodore. Based in England during the early part of the conflict he rose to air vice marshal and on his discharge from the air force in May 1946, McNamara continued to live in England and served on the National Coal Board in London from 1947 to 1959. He died of heart failure on 2 November 1961 aged 67 and is remembered as Australia's lone air VC from World War 1.

**Gaza March 1917**

# Top-scoring AFC pilot (1918)

**Harry Cobby**

# Harry Cobby (1894-1955)

The clash of opposing aircraft over the Western Front battlefields saw the emergence of a modern aerial knight; the fighter ace, and both Germany and their Allied opponents were well served with national heroes. For Germany there was Manfred Richtofen, with 80 victories the highest scoring ace of the war; Great Britain with Edward Mannock (73) and France's Rene Fonck (75). Australia too had its aces, Captain Little (47) and Major Dallas (39), although both pilots were serving with the Royal Naval Air Service. Captain Harry Cobby was unique in that he was top-scoring Australian Flying Corps pilot with 29 victories.

Arthur Henry Cobby was born at Melbourne in 1894 and on completion of his secondary education in 1912 was commissioned into a local militia unit. With the declaration of war in 1914 he attempted to enlist in the AIF, but for a period his employer refused to release him until 1916 when he managed to join the AFC, becoming a founding member of No.4 Squadron. The unit embarked for England in January 1917, arriving March 1917 to undergo further training prior to the move to France in December.

Based in the Pas-de-Calais area No.4 supported Allied forces during the German spring offensive, which was launched in March 1918. On 21 March Cobby was credited with two Albatros DVs, which was the day Manfred Richtofen was brought down by ground fire over the Australian lines. Cobby's abilities were recognized with his appointment as flight commander and promotion to captain.

Meanwhile he added two more to his tally, one being an observation balloon, which were considered as dangerous and valuable targets. On 3 June Cobby was recommended for the Military Cross, however this was changed to a Distinguished Flying Cross, and following his shooting down of three German aircraft on 28 June he was awarded a bar to the DFC.

With his score currently at fifteen, Cobby and a fellow pilot engaged five Pfalz scouts, resulting in Cobby downing two and his companion one. This action resulted in a second bar to his DFC, with favourable comments on his courage and brilliant flying. On 16 August Cobby led a bombing raid against the German airfield at Haubourdin, resulting in 37 enemy aircraft being destroyed. The following day he led a similar attack on Lomme airfield and as a result was recommended for a DSO.

With such results No.4 was recognized as the most successful fighter squadron in France, accounting for 220 victories and due in no small part to the leadership of Harry Cobby. In September he was taken off operations and transferred to a training unit in England, which was not to his liking, as he found instructing more stressful than combat. His applications to return to France were refused, resulting in Cobby taking no further part in operational flying. His final tally was 24 aircraft and 5 balloons, making him the top-scoring AFC pilot.

Cobby returned to Australia in 1919 and was one of the original 21 officers on the strength of the newly-raised Royal Australian Air Force in 1921. As a career officer he rose from flight lieutenant in 1921 to wing commander in 1933. In 1936 he retired from the air force to join the Civil Aviation Board, and at the outbreak of World War 2 he re-joined the air force. Promoted to group captain, Cobby held various commands at bases around Australia, and in 1946 was officially discharged from the air force. From there he re-joined the Department of Civil Aviation and served as Regional Director (NSW) from 1947 to 1954. In November 1955 Cobby collapsed at his office and died later that day, an iconic member of the AFC and RAAF.

**Sopwith Camel**

# First England-Australia flight (1919)

**Ross Smith**

# First England-Australia Flight 1919

Ross Smith gained fame as a decorated airman and pioneering trail blazer. Born at Adelaide 1892, the elder of two brothers, Ross and Keith Smith were educated at Queen's School Adelaide and later at a boarding school at Moffat, Scotland, their father's birthplace. On his return to Australia Ross became a member of a local militia unit and on the declaration of war in 1914 he enlisted as a private in the 3rd Light Horse Regiment, AIF, and in October he was promoted sergeant.

Ross served at Gallipoli from May 1915 to October, when he was invalided to England due to sickness. Returning to his unit in April 1916 Lt. Ross Smith saw action at the Battle of Romani and in July 1917 he transferred to the AFC. As an observer and later a pilot he saw extensive service in Palestine with No.1 squadron AFC. By the end of the war he had been awarded the Military Cross and Bar, the Distinguished Flying Cross and two Bars, plus the Air Force Cross for non-operational flying. This was for extensive long-distance flights in the Handley Page 0/400 bomber, which formed part of the squadron's equipment.

By contrast, Keith Smith's flying career saw no combat, despite his enlistment in the Royal Flying Corps and subsequent posting to No.58 squadron, a bombing unit which left for France in January 1918 but saw no active service.

However the Smith brother's careers were about to enter a new and historic phase, when in March 1919 the Australian Government offered a substantial prize of £10,000 for the first Australian crew to fly from England to Australia in under thirty days. At the time many ex-wartime pilots were awaiting repatriation to Australia, which encouraged a number of them to enter the race.

For the average airman the logistics of the event were quite daunting, where in most instances his longest flight would rarely have exceeded 90 kms. Navigational skills were generally basic, and few had the ability to fly in heavy cloud. Facing the aspiring race entrant was a flight of 19,000 km. over largely uncharted territory and also involving long ocean hops. Perhaps the greatest handicap was a lack of suitable machinery. Although there was a glut of war surplus aircraft available; these were mainly single-engine types, which certainly did not inspire confidence when facing an ocean crossing.

In early 1918 the Blackburn Company had produced a twin-engine biplane for anti-submarine operations, named the Kangaroo. This ungainly type saw limited service until the November Armistice, but it did offer the range to be a contender in the race. A four-man Australian crew made a successful start to the race but ran into trouble over the Mediterranean; and were lucky to escape with their lives when they force-landed on the island of Corfu. At least the Kangaroo had achieved a greater distance than the previous three starters, all of whom crashed, two of them fatally. Realistically the only starters likely to succeed were brothers Ross and Keith Smith, with Ross Smith's Middle East experiences later proving invaluable during the race. Two Australian mechanics, Sergeants Wally Shiers and James Bennett completed the crew list.

A converted Vimy bomber was provided by the Vickers Company at Weybridge in Surrey for the epic flight; carrying civil registration, GE-AOU. In typical Aussie fashion it prompted the crew to transpose it to; 'God 'elp all of us'. In frigid conditions the intrepid crew took off on 12 November from Hounslow in Middlesex; in the midst of a bitter European winter, which saw them battling snow and gale force winds across France to their first set-down at Lyon. Strong headwinds on the next leg to Rome forced them to land at Pisa where heavy rain had turned the airfield into a boggy morass. For a day and a half the Vimy was trapped there before it could be extricated. Despite further rain they pressed on to Rome, Taranto, Greece and the island of Crete. During their stay at Crete the aircraft received a thorough overhaul before the long sea crossing to Cairo; this was achieved without incident. In their passage through Mesopotamia and Arabia, Ross Smith's wartime experience enabled them to make up for time lost at Pisa.

Following a safe negotiation of the Arabian Gulf and inhospitable terrain of Persia a weary crew made a landfall at Karachi in western India. At Delhi they took a brief rest to recover from the strain of the previous stages. Through Burma and down the Malay Peninsula the flight was punctuated by tropical storms that forced them to fly blind on the primitive instruments of the time. At Singapore they were obliged to put down on a rain-sodden racecourse, the only landing area available. By that stage 22 days had elapsed since their departure from Hounslow; and they still faced a further 4000 kms, involving a passage over ocean, jungle and mountains. At Surabaya they landed at a newly constructed airfield and once again found themselves hopelessly bogged.

Keith Smith suggested that a roadway of bamboo mats might enable them to take off for the final legs to Bima and Darwin. An army of workers was enlisted to lay the mats and after 24 hours trapped in Surabaya they were able to take off for Bima; their departure point for the final obstacle; the Arafura Sea.

Facing the crew on their final stage was a five-hour flight; with a degree of security being provided by the cruiser HMAS Sydney, which was dispatched from Darwin to patrol the flight route should they be forced down. No problems were experienced and at 3pm December 10 they landed at Darwin, and were declared the winners in a time of 27 days and 20 hrs. For their determination and skill in completing the epic flight Ross and Keith Smith received knighthoods; while Bennett and Shiers were commissioned. It is pleasing to record that the historic Vimy still exists in honourable retirement on permanent display at Adelaide International Airport.

Of the other entrants the only crew to finish was Parer and MacIntosh in a single-engine DH9. They overcame a multitude of problems and eventually arrived in Australia seven months after the Smith brothers. Their historic DH9 was the subject of a recent restoration and is in storage in Canberra ACT.

Sadly, Ross Smith was killed in 1922 when his Vickers Viking amphibian crashed at Brooklands aerodrome in heavy fog. At the time the Smith brothers were planning a round the world flight, and on that fateful day Keith Smith was delayed for a test flight with his brother. Ross chose to ignore the foggy conditions and paid the penalty along with James Bennett, their staunch companion on the Vimy flight. Keith Smith retained his interest in aviation and in later years he became an executive with Qantas airlines until his death in Sydney in 1955.

**Departure from Hounslow 12 November 1919**

# Australia's Lone Eagle

**Bert Hinkler**

# Bert Hinkler 1892-1933

Born in 1892 at Bundaberg, central Queensland Herbert John Hinkler was the quiet achiever of Australian aviation. From an early age he displayed an interest in aeronautics and by the time he was twenty the diminutive Hinkler had built and flown a series of man-carrying gliders. In order to advance his aviation career he travelled to England and during World War 1 he served with the Royal Naval Air Service on the Italian and Western Fronts.

Following the Armistice of 1918 Hinkler settled in England and eventually he became test pilot for A.V. Roe, a leading aircraft manufacturer of the time. He was an interested observer of Ross and Keith Smith's 1919 England- Australia flight and was determined to emulate their feat but in a totally different manner. Instead of a large aircraft and elaborate preparation he opted for a miniature Baby Avro, a basic monoplane design powered by a modest 35hp engine, and could be regarded as the genesis of the future light aeroplane.

Despite official discouragement Hinkler set off from Croydon on 31 May 1920. He reached Turin non-stop in ten hours, which was a world record for a light aircraft. In Iraq he became victim to officialdom when authorities there refused him permission to fly over the Iraq desert. Hinkler's flight had officially failed but at least he had demonstrated the practicalities of light aircraft.

Later in the year he shipped the Avro to Australia where he made a flight from Sydney to Bundaberg, this was a significant event, for he was able to fulfil a promise to his mother that one day he would fly home in his own aeroplane. He returned to England and continued his career as chief test pilot for Avro and in 1927 the Avro Avian, a 2-seat biplane of metal construction and fabric covering went into production. Powered by an 80hp Cirrus engine, the Avian was capable of flying over 1,000 miles non-stop; as Hinkler demonstrated when he flew from London to Riga, a distance of 1,200 miles.

On 7 February 1928 he embarked on his record-breaking flight to Australia; a great challenge as it represented the first realistic solo attempt. His first stop was Rome, one of many he made through extremes of bitter cold and merciless heat. He crossed the Mediterranean Sea to Tobruk and then on to Basra and the forbidding Arabian Desert. Hinkler reached Karachi on 14 February, setting up a new record from England to India.

Apart from a minor repair to an oil tank, the Avian was performing faultlessly. He completed the crossing of the Indian continent and on the next stages to Singapore and Java he experienced conditions of heavy rain and poor visibility. Possibly the most hazardous stage was the final one from Bima, an island east of Java; from where he was faced with a 600 miles crossing of the lonely Timor Sea. There were no dramas and at 8pm on 22 February Hinkler landed at Darwin, fifteen and a half days after leaving England. His actual flying time was 134 hours or 5 days and 14 hours.

The quiet achiever had made the first solo flight from England and the first in a light aeroplane. Awards and acclaim were bestowed on the unassuming Hinkler. These ranged from a triumphal passage around Australia, an honorary rank of squadron leader in the RAAF to an award of the Air Force Cross from H.M King George V. Hinkler's record stood intact until May 1930 when Charles Kingsford Smith lowered the time to nine days and 22 hours. His machine was also an Avian, which he named 'Southern Cross Junior'.

In 1931 Hinkler set out on the flight of his career, a journey from New York to London. His machine was a DH Puss Moth powered by a 130hp Gypsy Major engine, and while it did not possess the range to achieve a crossing of the North Atlantic he achieved his goal with a novel approach to the problem. The Puss Moth also lacked certain equipment required for night flying over The United States, making it necessary to fly direct to Bermuda, an 18-hour flight and the first between the two cities. From there he flew south in easy stages from Venezuela to Port Natal on the Brazilian coast.

He was then faced with a 3,200 km ocean crossing to Bathurst on the west coast of Africa. Six hours into the flight Hinkler encountered fog and heavy rain, which persisted throughout the entire crossing. His aircraft was not equipped with blind-flying aids, forcing him to fly by compass and his own instincts; yet so accurate was his course that he made a landfall with an error of just one degree. That stage was the first west/east crossing of the South Atlantic and in the process he made the longest non-stop flight in a light aircraft. Hinkler reached London on 7 December 1931, having flown 16,000 km. to achieve his goal, and in a quite unique manner.

But the skies that highlighted Hinkler's exceptional abilities were about to betray him. In January 1933 he set out in the Puss Moth in an attempt on the England-Australia record. At the time it was held by C.W.A. Scott in a Gypsy Moth in eight days and 20 minutes. On the same day after leaving England Hinkler was reported over the Italian Alps in the vicinity of the Pratamagno

Mountains. Following that sighting there was no further news until 1 May when itinerant forestry workers discovered wreckage of the Puss Moth. Nearby was Hinkler's body.

Bert Hinkler was later buried in the village cemetery in Pratamagno village. His achievements were not forgotten, in particular in Bundaberg where the original family home has been set up as a museum to honour his memory; also in a Brisbane museum where the trusty Avian is on permanent display.

**DH Puss Moth 1933**

# First Pacific crossing 1928

**Charles Kingsford Smith**

## Kingsford Smith and the 1928 Pacific flight

Australian aviation produced more than its fair share of heroes and trail-blazers, but of them all, one stands head and shoulders above the pack. He was Charles Kingsford Smith whose achievements were legendary and far too many to note within these pages; however one particular flight eclipses all others, which the author elected to record, Smithy's 1928 Pacific crossing.

Charles Smith was born in Hamilton, Brisbane in 1897, the youngest in a family of six siblings, four brothers and two sisters. The Smith family was an itinerant one, rarely staying long at the one address, and one of these moves was to Canada in 1903. While there, the title 'Kingsford' was added to the family name, which was Mrs. Catherine Smith's maiden name and added a distinctive tone to the more common 'Smith', while it also made life easier for the postmen in a street populated with 'Smiths'.

The Canadian venture was not a successful one and after a generally miserable period, by 1908 the family had drifted back to Australia. They settled in Sydney, but not in permanent manner, moving from one address to another. By 1914 Charles had completed his secondary studies and was serving an apprenticeship as an electrical engineer. The prospect of going to war greatly appealed to Smithy and with his parent's approval he enlisted in the AIF on 9 February 1915, the date of his 18th birthday.

His unit was the 4th Signals Group of the 4th Divisional Signal Company, and as Private Kingsford Smith he went on to serve as a despatch runner at the ill-starred Gallipoli campaign. By December 1915 the disastrously failed expedition ended in evacuation and the survivors were back in Egypt for reinforcement and recuperation. In April 1916 the five AIF divisions were transferred to France and were quickly pitched into battle alongside the British Expeditionary Force.

During a four month period Sgt. Smith served as a motorcycle despatch rider, and surviving that hectic time, in September 1916 he applied for a transfer to the Royal Flying Corps. In March 1917 he was commissioned lieutenant and began pilot training, a dangerous period when one pupil in three was failed, seriously injured or killed. With barely thirty hours flying time Smithy was awarded his wings and posted to No.23 squadron, based at La Louvie in the Pas de Calais area.

His unit flew the French Spad VII, which was better suited to ground attack missions rather than air-to-air combat. Nevertheless, in a period of six weeks Smithy was credited with five confirmed victories, thus elevating him into the 'ace' category. On 14 August 1917 Smithy became a victim himself, when on a dawn patrol he was jumped from above and severely wounded in his left foot. Between moments of swooning in agony and consciousness he managed to make it home to base and survived a heavy landing, in which his bullet-lacerated Spad was virtually a write-off. Smithy's wounding was a serious one and for the moment he was taken off combat and never flew operationally again.

Post-war he spent time in America, performing dare-devil aerial stunts for the movie cameras, prior to returning to returning to Australia, five years since the heady day of enlistment in 1915. Smithy's piloting skills enabled him to find employment with West Australian Airways for a one-year period, and while there he struck up an unlikely friendship with a fellow pilot, Keith Anderson, who like Smithy saw combat on the Western Front. Anderson's taciturn personality was precisely the opposite of the effervescent Smithy, yet the duo was later to create a successful motor haulage business. With profits gained from the enterprise they purchased two elderly Bristol Tourers and in 1927 moved to Sydney with the intention of creating local air charter work.

The partner's aims were focussed on an eventual aerial Pacific crossing, which at the time would seem to be wildly impractical. The proposed air charter venture was sadly unsuccessful; however a meeting occurred with a third party, which had a profound effect on their fortunes. Charles Ulm had no piloting skills; instead his talents lay in organizing, which he pursued in a ruthless, unwavering manner. Anderson was of an easy-going nature and soon found himself relegated to the background by the thrusting, single-minded Ulm.

In an attempt to gain sponsorship for the Pacific flight, Smithy and Ulm embarked on a round-Australia flight using one of Smithy's Bristol Tourers. Their circumnavigation was completed in a record 10 days, which created favourable responses from various sources. Confident that the financial problems were covered, the trio boarded a steamer for The United States in July 1927 to prepare for the Pacific flight.

Shortly after their arrival they were able to purchase an aircraft suitable for their requirements from Polar explorer Hubert Wilkins. This was a Fokker FV11 B tri-motor, one of a pair that Wilkins had used on a recent Polar flight. Unfortunately the Fokker; which they named 'Southern Cross' came without motors and instruments. These were beyond their current funds, but with an additional £1,000 from the NSW Government and a further £1,500 from Melbourne businessman, Sydney Myer the project gained momentum.

'Southern Cross' was then fitted out with three Wright Whirlwind motors, instruments and long-range tanks, but the partners' money was fast running out, and with the situation becoming desperate, Anderson returned to Australia, convinced that the project had no future. Smithy and Ulm were basically destitute and burdened with debts that amounted to $16,000. Reluctantly they made the decision to sell the Fokker and return to Sydney

In a stroke of good fortune their financial problems were solved by the intervention of millionaire banker, Alan Hancock. This quiet businessman was also an experienced marine navigator with a great interest in aviation. He was sufficiently impressed with the Australians to take over their sizeable debts and fund the Pacific flight. Two Americans were recruited to complete the crew; Harry Lyon, former ship's captain and a superb navigator, and James Warner, a radio operator with many years service with the U.S. Navy and Merchant Marine.

Here was Smithy's moment of destiny. For years his obsession was the Pacific flight, and after incredible setbacks it was about to be realized; however it also carried a bitter irony for Anderson; being denied his opportunity to be part of the team.

On the last day of May 1928 the grossly over-laden 'Southern Cross' lifted off from San Francisco's Oakland airport. Their take-off was the most critical stage of the flight, so laden with fuel their speed was barely above the stall, and was a situation they faced for several hours until enough fuel was burned for them to gain precious altitude. Communication between navigator and pilot was achieved by way of notes pushed through a tube that ran from rear cabin to cockpit. Engine noise was so intrusive that conversation between pilot and co-pilot was impossible. They too were compelled to resort to scribbled notes to each other.

For the first stage of the journey they were blessed with ideal weather conditions; clear and sunny and with the bonus of a tail wind. Twenty-four hours into the flight they were faced with a crisis when the batteries for the receiver and transmitter expired, making it absolutely vital that Lyon's navigation was faultless. Several times land was sighted, but to their disappointment the sightings proved to be cloud masses. In the midst of their problems Warner received a reassuring signal from a shore station giving them an indication of their position. At the same time Lyon managed to obtain an accurate sun shot that verified the signal.

For another hour 'Southern Cross' cruised above heavy cloud while the crew watched anxiously for a long-overdue sighting of land, with the situation becoming a race against the possibility of running out of fuel and a landfall. At last they were rewarded with the sight of the 14,000 feet volcano, Mauna Kea on Hawaii Island; Lyon's navigation had been impeccable. Sixty minutes later they landed at Wheeler Field at Honolulu to complete the 2,300-mile crossing in 27 hours 30 minutes.

Following the euphoria of the first leg, 'Southern Cross' and its crew faced the most critical stage of the journey; 3,150 miles to Fiji and the first ever attempt by air. Smithy chose a beach at Barking Sands on the outer island of Kauai for the take-off. In retrospect it bordered on the suicidal with the all-up weight dangerously close to the aircraft's limits, making it a nerve-wracking experience for its crew and for the spectators. They watched in trepidation as 'Southern Cross' staggered into the air with its landing wheels skimming the waves as Smithy kept it airborne and it was some time before they clawed their way to an altitude of 500 feet and received assistance from a helpful tail wind. Three hours into the flight the troublesome radio problems returned, leaving them with only the capacity to transmit messages. Around mid-day they hit the inter-tropical convergence zone, an area that was host to storms of extreme ferocity.

Menacing clouds exploded with thunder and lightning, rising to altitudes of 35,000 feet; and making it quite impossible to fly over. 'Southern Cross' plunged into the boiling mass where it was flung about like a leaf in a storm.

All of Smithy's skills were needed to keep it airborne, in conditions that were akin to flying through a waterfall that drenched the pilots from head to foot. At one stage the aircraft was almost flipped upside down, so severe was the turbulence. Eventually they emerged into clear skies, and by Lyon's reckoning they had covered over 1,000 miles, almost one third of the distance. To their horror they saw ahead of them another threatening cloud mass, hundreds of miles across. Once more it became a desperate battle against the elements as Smithy endeavoured to climb above the worst of it; and after three terrifying hours they emerged into a moonlit starry night. According to Lyon they were 1,100 miles from Fiji and with 500 gallons of fuel remaining.

By Smithy's calculations that was inadequate, which meant they might have to find an alternative island for a set-down. The remaining hours until daylight dragged by in an atmosphere of extreme tension and when the time came to hand pump the remaining fuel they discovered there was ample to cover the distance. Lyon was able to obtain a shot of the sun during the morning, which showed they were north of their intended course. It was a tribute to his navigation after the constant detours during those violent storms and without a single radio bearing from ship or shore to assist him.

'Southern Cross' arrived over Suva in the early afternoon and Smithy was not impressed with the miniscule landing ground at Albert Park. He made a search for a suitable beach as an alternative but they were even less promising; it had to be Albert Park. An enormous crowd had gathered to witness their arrival and they were privileged to see Smithy perform a masterly landing in the confined space. Just when it seemed 'Southern Cross' was about to career into some large trees Smithy swung away in a spectacular ground loop. The Americans were crouched in the rear of the fuselage in an endeavour to hold down the tail of the aircraft, and during the violent landing the unfortunate Warner was flung through the fabric and knocked unconscious. He was fortunate not to be run over, and after being rendered first aid he quickly recovered.

For the final leg to Brisbane Smithy used Naselau Beach from where they took off on Friday 8 June. Being a far shorter flight; a maximum fuel load was deemed unnecessary, and as they headed into a brilliant clear sky, crewmembers were quietly confident they were almost home. With the difficult part of the journey behind them they could hardly miss a target the size of Australia. Their euphoria was dealt a severe blow when soon after nightfall they flew into an electrical storm of terrifying proportions. All of Smithy's blind- flying skills were called on to keep 'Southern Cross' airborne. In an attempt to find a more favourable situation they climbed to 9,000 feet, where they faced bitterly cold conditions. Even at that altitude the Fokker was forced up and down in alarming surges. In the rudimentary cockpit the pilots were flung out of their seats as they clung desperately to the control wheels. For the Americans they fared even worse in conditions that at times rendered them weightless.

During the five hours of their ordeal it was impossible for Lyon to navigate and when they did emerge from the storm he estimated they were at least 100 miles off course, making it fortunate their goal was Australia and not a speck in the ocean like Fiji. Around 8am a shadow appeared on the horizon, which according to Lyon was the coast of northern New South Wales. Smithy identified it as Ballina, 110 miles south of Brisbane. An error of that magnitude on the Fiji leg did not bear consideration.

As they neared Brisbane, 'Southern Cross' was welcomed by a flight of light aircraft and at 10.15 a.m. Smithy made a textbook landing at Eagle Farm aerodrome. It would be difficult to put a precise number to the crowd that had waited since 3am. That figure varied from 15,000 to 40,000 but there was no doubting their adulation for the crew. Smithy became the centre of attraction as women attempted to smother him with kisses while men fought to shake his hand, heralding the beginning of an era of pop-idol fame for the effervescent Aussie.

Their flight statistics showed they had covered 7,220 miles in eight and a half days and in a flying time of 83 hours. Present day travellers might take a moment to compare their journey in the pressurized ease of a 747 Jumbo to the acute discomfort and appalling danger of that Pacific crossing of 1928.

**Southern Cross May/June 1928**

# The Kookaburra tragedy (1929)

**Keith Anderson**

# Kookaburra

Pioneer aviator and World War 1 fighter pilot Keith Anderson was closely following developments in the search for Charles Kingsford Smith and crewmembers. Radio contact with 'Southern Cross' was lost when they encountered violent dust storms and made a forced landing on mud flats in northern Australia. This was March 1929, ten months after Smithy's historic flight from California to Australia. They became lost during the first stage of a flight to England; the purpose of which was to purchase new aircraft and to also complete a round the world flight in the same aircraft.

For Anderson the entire situation carried a bitter irony. The original planning for the Pacific flight was formulated between himself and Kingsford Smith. In the early 1920s they worked together in Western Australia in a variety of projects. An aerial taxi service proved unrewarding with its frequent forced landings; however a truck haulage business enabled them to accumulate enough funds to move to Sydney.

A fortuitous meeting with Charles Ulm convinced Smithy that here was the person they needed to handle the business side of their activities. Ulm's forceful attitude tended to push the taciturn Anderson into the background; with events reaching a climax when the trio was in America seeking backers for their Pacific flight.

Anderson was convinced the project had no future and returned to Australia. He could only watch in dismay when their financial problems were solved and the Pacific crossing achieved in May 1928. In a surprising and generous gesture Smithy presented Anderson with a cheque for £1,000 which was a considerable sum of money in those days. With those funds he purchased a Westland Widgeon, a small parasol monoplane powered by a 60hp engine, which was not a practical choice, but with 'Kookaburra', as Anderson named his new acquisition he planned to make record-breaking flights for a light aircraft.

On two occasions in their Western Australian partnership Anderson made forced landings in the harsh desert and on both occasions his rescuer was Kingsford Smith. Perhaps he saw Smithy's disappearance as an opportunity to repay the gesture. Another factor would be favourable publicity if his rescue attempt was successful. Anderson contacted Bob Hitchcock, a resourceful mechanic as his companion on the venture, and with the barest of preparation they set off for Central Australia. Realistically the little Widgeon was ill-suited for such a flight, with its modest horse-power, lack of radio and worst of all, a compass that was known to be inaccurate. The ill-prepared duo reached Alice Springs after a battle with headwinds, dust storms and a forced landing with engine trouble en route.

On the following morning they set out in the grossly overloaded Widgeon. Their intention was to follow the overland telegraph to Woodford Crossing, and it was at that stage that Anderson made an unaccountable decision to save time by diverting cross-country despite the faulty compass.

Shortly afterwards things began to go terribly wrong. Once again the fractious engine lost power and left them with no alternative than to make another forced landing, this time in the inhospitable Tanami Desert. Repairs to the troublesome engine were soon affected but they faced a hopeless situation to get airborne from the sandy surface, overgrown with tangled scrub. Their hasty preparations became woefully obvious; with a lack of tools and the barest supply of food and water.

With just a penknife and their bare hands the hapless pair endeavoured to make a clearing for a take-off. All their labours were to no avail when the tyres of the landing wheels punctured on the harsh turpentine scrub. Their future was bleak indeed. Unless help came with a rescue party they must surely perish through thirst and exhaustion.

On April 12, two days after 'Kookaburra's' disappearance a search plane located 'Southern Cross' on mud flats on the north west coast of the Northern Territory. All four crewmembers were alive, but suffering with exhaustion from their experience. The air search then concentrated on 'Kookaburra', and on April 20 there was a sighting by a Qantas aircraft. One person was seen lying beneath a wing but of survivors there appeared to be none. Due to the dense scrub it was impossible for the pilot to land, but he was able to relay their position by radio.

Following the aircraft's return to Wave Hill station, a ground party was dispatched; and after an arduous trek the horsemen and black-trackers reached the site on April 27. The body under the wing proved to be Hitchcock; while a further search located Anderson's body some distance away. Temporary graves were dug near the aircraft until more permanent arrangements were made in their home states. Before he succumbed to the conditions Anderson kept a record of their experience pencilled on the aircraft's rudder. That poignant scrap of fabric was retrieved and can be seen in a Perth museum.

Another land party was organized, using a six-wheeled lorry capable of traversing the sandy wastes. That expedition was an epic in itself and after a variety of setbacks it reached the site some weeks later. The grim task of exhuming the bodies and returning them to their native states was then carried out; Anderson's was to Sydney and Hitchcock's to Perth

Hitchcock's funeral was a private affair, attended by family and close friends. In an altruistic gesture the Commonwealth Government gave an assurance to his widow that a service career would be found for her three sons when they became of age. Anderson was accorded a full military funeral after his body lay in state in St. Stephen's Presbyterian Church in the city. His burial took place at a cemetery on Sydney's North Shore and was attended by thousands of mourners, An evocative touch was provided by a flight of Gypsy Moths as they flew overhead in the formation of a cross, making it ironic that Keith Anderson's moment of glory came in the solemnity of his burial

This sad chapter of events had unpleasant repercussions for Kingsford Smith and his crew. They were all called before a Royal Commission to explain the circumstances leading up to their disappearance. Damning allegations were made that the whole affair was a publicity stunt, and also to the inadequate flight preparations. They emerged from the enquiry cleared of the charges but suitably chastened by the experience. Smithy and Ulm's reputations were publicly tarnished and they never regained the absolute hero status they enjoyed following the 1928 Pacific flight.

**Kookaburra March 1929**

# Southern Cloud loss (1931)

**T.W. Shortridge**

# Southern Cloud 1931

One of the most enduring mysteries of early Australian aviation began on 1 March 1931. 'Southern Cloud' was flagship of Australian National Airlines, the company that Charles Kingsford Smith and Charles Ulm launched on New Year's Day 1930. Their fleet initially comprised three Avro 10 tri-motors, 'Southern Cloud' (VH-UMF), 'Southern Star' (VH-UMG) and 'Southern Sky' (VH-UMH). Eventually they were joined by 'Southern Moon' and 'Southern Sun'. The Avro Ten was basically an Anglicised version of the proven Fokker FV11B series and five examples were supplied to ANA.

The new aircraft were flown by experienced captains and maintained by ANA's efficient ground engineers. These factors resulted in a service that gained an enviable record for reliability and safety. A daily service was available from Sydney to Brisbane, Sydney to Melbourne and at a later date from Melbourne to Launceston. One situation that was causing concern was the absence of on-board radio but experiments were being carried out despite a scarcity of ground stations. Consequently once an aircraft was committed to a flight no means were available to warn the crew of deteriorating weather conditions ahead of them.

Captain T.W. Shortridge faced such a situation when he departed Sydney's Mascot aerodrome on Saturday 1 March 1931 on a scheduled flight to Melbourne. On board 'Southern Cloud' were six passengers and two flight crew. Eight passengers were originally booked, but two chose to postpone their flight until the following Monday. Early forecasts from Sydney Weather Bureau were for gale force winds and heavy rain along the route; however that was only relevant up to 9am and did not predict the extreme conditions experienced in southern New South Wales and Victoria.

Captain Shortridge was aware of what to expect; a bumpy and slow passage and hopefully no worse than what one encountered on previous flights. Along the route the aircraft was sighted over Moss Vale and was later heard over Goulburn. Two hours into the flight, the Melbourne Weather Bureau warned ANA of the severity of the conditions, and one can only imagine their alarm on receiving the news and being powerless to inform Shortridge of the situation.

By late afternoon their fears were justified; with the apparent disappearance of 'Southern Cloud'. With its non-arrival in Melbourne there was genuine concern that it had come down in an uninhabited area of the Snowy Mountains. On the following morning a comprehensive search was initiated, involving the ANA fleet and other civil aircraft, plus units from the RAAF. Main focus of the search was on the rugged Snowy Mountains where stockmen claimed to have sighted or heard an aircraft in low cloud in that area. All these sightings proved negative and after two weeks of intensive effort the search was reluctantly abandoned.

An official enquiry concluded that 'Southern Cloud' encountered extreme conditions of thunderstorms, rain and heavy cloud. In such conditions it was likely that the pilot became lost and was unable to recover from the situation. Smithy and Ulm were dealt a body blow with the adverse publicity resulting from the affair, plus the costs of the search. Other factors were also against them; with the effects of the Great Depression beginning to bite into people's lives. Air travel was still regarded as something of a luxury, and merely the province of the wealthy.

Another factor with ANA was that it received no Government subsidies, unlike its main competitor, Qantas. In order to lift ANA's profile and gain lost prestige they took the bold step to fly the Christmas airmail to England and return. 'Southern Sun', piloted by Scotty Allen departed Melbourne in late November on the first leg of the historic flight. Disaster struck once more when the aircraft was destroyed on take-off from a waterlogged Malayan airfield. 'Southern Star', piloted by Kingsford Smith was hastily dispatched to recover the mail and after a hectic 13 day flight Smithy delivered the mail to London's Croydon aerodrome.

This was a convincing demonstration of their ability to deliver the goods, but even that was insufficient to salvage ANA's diminishing fortunes. Finally it was time to wind up the operation and close their doors. Only for the tragic loss of 'Southern Cloud' and the demise of ANA, it is reasonable to assume that the Australian airline industry today would be composed of vastly different operators.

The mystery of the lost airliner remained unsolved until October 1958. In 1949 work began on the vast Snowy Mountains Hydro Electric Scheme. While on a bush walk in October 1958 a workman chanced upon some wreckage in the remote Tooma River Gorge and was clearly the remains of an aircraft embedded into the hillside. Removal of manufacturers plates revealed that it was 'Southern Cloud'; missing since 1931, and judging by the depths to which the motors were buried it appeared to have flown into the heavily wooded hillside at full power.

Skeletal remains and personal effects were discovered amongst the wreckage. One of these items was a wristwatch that had stopped at 1.15; and if the watch had stopped on impact it appeared that 'Southern Cloud' had taken over five hours to cover less than 200 miles. Captain Shortridge was possibly of the opinion that he was well clear of the mountains and began what he thought was a descent towards Melbourne. Instead he crashed to destruction in the loneliest of areas.

**Southern Cloud March 1931**

# Stella Australis 1934

**Charles Ulm**

# Charles Ulm (1898-1934)

Future airline operator Charles Ulm was born in Melbourne in 1898, and at the outbreak of war in 1914 he enlisted in the A.I.F. despite being under age. He served at Gallipoli where he was wounded and subsequently repatriated to Australia. When authorities became aware of his age he was discharged from the army, but Ulm Senior was vehemently anti-German and urged Charles to re-enlist once he turned 18, and for a second time he became a member of the A.I.F, going on to serve on the Western Front. Again he was wounded and whilst recuperating in England he developed an interest in aviation; enough to influence him to eventually embark on an aviation career.

His early business ventures were dogged by failure but a meeting with Charles Kingsford Smith put the seal on a famous partnership. Ulm, with his organizing skills was the perfect foil for Smithy, the impulsive trailblazer. Smithy's original partner was Keith Anderson, like himself a pilot in World War 1. It would be difficult to find a more diverse pair; the extroverted swashbuckling Smithy and the gentle giant, Anderson; yet such was Ulm's influence in the new arrangement that Anderson found himself relegated to being a virtual outsider.

To gain sponsorship for their Pacific project Ulm organized a round-Australia flight in June 1927.; using one of Smithy's Bristol Tourers but pointedly it was Ulm and not Anderson who accompanied Smithy, despite Ulm's lack of piloting skills. Their time of 10 days shattered the previous record by 13 days. Ulm had been diligent in organizing favourable newspaper publicity for the project which brought generous financial returns from the press and also the State Government. Shortly after that flight the trio sailed for The United States, and with their available funds they were able to purchase a Fokker FV11B from Arctic explorer Hubert Wilkins. Melbourne businessman Sydney Myer provided additional funds for them to install new motors and instrumentation. But they still faced incredible problems before the Pacific flight became a reality.

Anderson was sufficiently disenchanted to withdraw from the partnership and return to Australia. Smithy and Ulm were totally down on their luck when a chance meeting with millionaire banker Alan Hancock completely altered their plans. Their benefactor cleared all their outstanding debts and financed the entire project; the Pacific dream became a reality. Two Americans, Harry Lyon and James Warner were enlisted to the team and in May 1928 they flew into history.

Smithy and Ulm became national heroes following that historic flight which demonstrated the possibilities of international air travel.

On New Year's Day 1930 Smithy and Ulm launched Australian National Airlines with facilities at Sydney, Melbourne and Brisbane, which took all of Ulm's organizing skills to create and maintain the operation. Air travel in those depression years was regarded as an unnecessary luxury, and the realm of the elite. Smithy's casual attitude to his airline was quite astounding. Occasionally he would captain a flight to Melbourne or Brisbane, but the monotony of those flights was something he preferred to avoid. His preference was to be making record-breaking flights and enjoying a gregarious life style.

ANA could not have survived for a day without the dynamic Ulm at the helm, but in March 1931 it was dealt a body blow with the loss of 'Southern Cloud', carrying 6 passengers and 2 flight crew. The cost of the search, plus a lack of public confidence in air travel led to insurmountable problems; and in late 1931 ANA was forced to close down its operation.

The partners went their separate ways; Smithy continued his trail-blazing flights while Ulm scraped out a precarious living with 'Southern Sun', which he purchased from ANA and re-named it 'Faith in Australia'. He employed the experienced Scotty Allan as pilot and between them they made several notable flights. In 1934 an opportunity arose for Ulm when tenders were called for an airmail service to the United Kingdom. All of his expertise went into the tender; and should it be accepted, would be a just reward for his earlier endeavours.

The other tender was submitted by Qantas, who had earlier entered into a joint venture with Imperial Airways, which meant there could only be one winner. Also, the tender was created around the employment of the new DH 86 coming into service. All of Ulm's approaches to De Havillands were met with a complete refusal to supply relevant details of the aircraft's specifications. In effect the whole affair was closed to Ulm and the contract awarded to Qantas. Ulm was justifiably disappointed with the outcome, but he was offered a crumb of hope with a future Government contract for an inter-city service in Australia.

With Allan's assistance he drew up information to demonstrate the feasibility of world flights; in particular the weather conditions for relevant areas. Ulm ordered a new twin-engine Envoy from the Airspeed firm in Portsmouth, who installed additional tankage for the long ocean hauls; however he chose to ignore Allan's advice about making the Pacific crossing from America in the March/April period.

Instead he insisted on leaving in January, which was totally unacceptable to Allan, who resigned from the venture. As a result, Ulm engaged Sydney-based George Littlejohn as co-pilot and Leon Skilling as navigator. Skilling's experience was limited to marine navigation; a much more forgiving area than aerial navigation, which requires a distinct 'feel' for the conditions. The Airspeed directors also questioned Ulm's insistence on the siting of the additional fuel tank, which was located between pilot and navigator positions, thereby isolating one another in the event of a crisis. Ulm remained adamant about the arrangement and test-flew the Envoy, which he named 'Stella Australis'. He expressed his satisfaction and then had it shipped to The United States.

In December 1934 they took off in the early evening from San Francisco for the first stage to Honolulu, with an expected arrival on the following morning. Somewhere into the flight they ran into difficulties and missed Honolulu. General consensus was that a stronger than predicted tail wind had pushed them west of Hawaii. One can only imagine the final desperate hours of 'Stella Australis', with its navigator isolated and unable to ascertain their position.

Their calls were heard throughout the morning and early afternoon; until the final message that they were about to ditch. An extensive air search failed to locate the aircraft or its crew who were listed as: *Lost at sea, presumed dead.* Charles Ulm's impatience and obstinacy had sown the seeds for a needless disaster and with it the loss of one of Australia's most enthusiastic pioneers.

**Stella Australis December 1934**

# Jubilee Air Mail 1935

**P.G. Taylor**

# Southern Cross and the 1935 Jubilee Mail

Those early pioneering years saw many famous partnerships; one of the most outstanding was that of Charles Kingsford Smith and Patrick Gordon Taylor. Both had flown in combat in World War 1 and as civil pilots in the early post war period. P.G. (Bill) Taylor pursued an aviation career during that period with a quiet fervour and with a particular crusade in navigation. Perhaps their greatest achievement was the first Australia-America flight. This was in October 1934 in a single-engine Lockheed Altair, 'Lady Southern Cross', and was regarded by many as Kingsford Smith's greatest flight; one that brought great acclaim but unfortunately for Smithy there were scant financial rewards.

Nineteen thirty-five was the year of the Silver Jubilee of H.M. King George V and Queen Mary. Part of the celebrations was a special Jubilee Mail flight to New Zealand, the contract for which had been awarded to Kingsford Smith. The original plan was for the flight to be performed by two aircraft; with Smithy at the controls of 'Southern Cross' and Scotty Allan in 'Faith in Australia'. Ever cautious, Allan expressed concern over the condition of the motors of 'Southern Cross', which were in need of a major overhaul. Another worrying factor was a recent welding repair to the exhaust manifold of its centre engine. When Allan relayed his findings to Smithy he was astounded at his indifference to the situation.

However their problems were not confined to 'Southern Cross'. 'Faith in Australia' was found to have serious leaks in its fuel tanks. It was all too much for the wily Allan who resigned from the operation, and Taylor became 'Faith's' pilot, but such was the extent of its problems that a decision was made not to use it. Instead, Taylor joined Smithy as co-pilot in 'Southern Cross' with John Stannage enlisted as wireless operator. One can only imagine Taylor's misgivings about the flight. The idea of heading out over the Tasman in a suspect aircraft was bordering on suicidal; but Smithy appeared unconcerned about their prospects; such was his absolute confidence in his faithful Fokker.

At 12.20am on 15 May 1935 they took off from RAAF Richmond. Shortly they crossed the coastline and began their Tasman crossing; a stretch of water that Smithy regarded as 'the world's worst bit of ocean'. Six hours into the flight and during Taylor's spell at the controls he became aware of an irregularity with the suspect exhaust manifold. He was indicating the problem to Smithy when suddenly the aircraft was seized in a fearful vibration, and the reason soon became frighteningly obvious. A section of manifold had broken away in the slipstream and smashed part of the propeller blade of the starboard motor. Smithy swiftly shut the motor down before it shook the aircraft to pieces.

Robbed of vital horsepower he increased power on the remaining motors to remain airborne. At that point they were less than halfway to New Zealand, prompting Smithy to head back to Australia and hopefully coax his stricken machine home. The situation developed into an exhausting struggle to keep 'Southern Cross' aloft in the crab-like attitude he was forced to maintain. Taylor estimated it would take 10 hours to reach Australia at their present speed; provided the overworked motors held out. To lighten their load Smithy ordered the crew to jettison excess fuel and equipment but refused to dump the precious mail. Hour after hour dragged on as Smithy and Taylor took their turn at the controls; and a worrying factor was oil consumption. Already the pressure was falling in the port motor, and unless replenished it would surely fail; that was the awful reality.

On his own initiative Taylor made the heroic decision to transfer oil from the lifeless starboard engine to the ailing port unit. Access to the outer motors was gained via a strut that linked to the engine mounting, which was a difficult enough task on the ground but Taylor was about to attempt the impossible. He removed his shoes, tied a light line around his waist and ventured out into the icy blast of the slipstream. By wedging his shoulders against the wing it was possible to make his way to the engine cowling and gain access to the oil reservoir.

First he would need to undo the large hexagonal sump plug. He inched back along the strut to reach for the shifting spanner that Stannage held out for him. Thankfully the nut yielded to his efforts, but what could he use for a container? He looked to the cabin and there was Stannage holding out a thermos flask. Once more he crept back along the strut and grasped the container. In spite of the slipstream he managed to trap some of the precious fluid. Then it was back and forth to the cabin to pass the thermos to Stannage who emptied the contents into a small suitcase. Finally he clambered back into the cabin, battered and exhausted from his ordeal; but the task was only half-completed.

Taylor was facing the ordeal of the blast of both motors in his efforts to replenish the ailing port motor. Smithy signalled to Taylor that he would temporarily stop the port motor while he topped up its oil supply. It was a desperate situation; with 'Southern Cross's' landing wheels literally skimming the water as Smithy re-started the port motor and regained some precious altitude.

But Taylor's incredible effort had retrieved the situation, if only temporarily. He hauled himself back inside and fell to the cabin floor, completely exhausted and gasping like a stranded fish. At that point Smithy gave the order to dump the precious mail. Perhaps he should have done it sooner but at least they were now able to gain some altitude.

Taylor's respite was only brief; for once more the flickering oil pressure gauge headed towards that dreaded zero, and they were still 200 miles from the haven of Australia. Wearily, Taylor hauled himself from the sanctuary of the cockpit and began the purgatory of another oil transfer.

Task completed; he collapsed on the cabin floor. With his oil-soaked clothing he gave the impression of some creature that emerged from the Black Lagoon. That he was able to retain his footing out there on that strut with hands and feet drenched with oil bordered on miraculous. Finally at 4pm. with its motors in their death throes 'Southern Cross' crossed the coast at Cronulla and limped into Mascot. They had been airborne for fifteen hours, the last nine of them in a situation of utter peril. The exhausted crew emerged from their nightmare; barely able to comprehend that they had survived.

Taylor's heroic feat was recognized with the award of the George Cross, the highest civilian decoration for bravery and it was also the last great flight for Smithy's 'Old Bus'; for shortly afterwards the Commonwealth Government agreed to purchase the historic aircraft for the sum of £3,000. Smithy delivered it to RAAF Richmond and following an emotional handing-over ceremony it found a new home. That Smithy never saw the promised money for 'Southern Cross' was a national disgrace, despite his pleas for settlement. Not until he disappeared in November 1935 on that fatal England/Australia flight did the politicians shake off their lethargy and release the funds to Smithy's widow.

The iconic aircraft eventually found a permanent home at Brisbane international airport, the site of its arrival at Eagle Farm in 1928 and can be viewed by all and sundry in its glass-encased hangar.

**Southern Cross May 1935**

# Lady Southern Cross 1935

## Charles Kingsford Smith

# Lady Southern Cross - Smithy's Last Flight

The 1934 MacRobertson Air Race attracted worldwide interest in aviation circles, however Charles Kingsford Smith's plans to compete with a British aircraft were thwarted by the politics of the time when De Havillands informed him they were able to provide one of their new Comet racers but not with variable pitch propellers. Without that refinement his machine would be totally uncompetitive.

Frustrated, he sailed for America and purchased a second-hand Lockheed Altair. He now had a potential winner; in theory at least. The Altair was powered by a 550 hp Wasp engine with variable pitch airscrew, which gave a cruising speed of 180 mph. With its retractable undercarriage this was an advanced feature also, but the Altair was handicapped by a standard fuel capacity of 150 gallons. Smithy overcame the problem by increasing the tankage to 400 gallons; thereby invalidating its U.S. certificate of airworthiness.

Smithy's casual attitude to regulations was legendary, and despite warnings from Australian Civil Aviation to obtain U.S. approval he ignored them and had the Altair shipped to Australia. On arrival in Sydney he was informed by Customs that the Altair was a prohibited import and was barred from flying in that country.

Another contentious issue was the name 'Anzac' he had painted on the aircraft, despite the fact that Smithy was an Anzac himself. To appease the Government he changed it to 'Lady Southern Cross'; one more variation on a theme which began with the original 'Southern Cross'.

As a concession he was permitted to unload the Altair onto a barge and ferry it to Anderson Park on Sydney's North Shore, and from there he made a masterly take-off to Mascot aerodrome. In an attempt to placate the race organizers and the Royal Aero Club in London Smithy appealed to Australian Civil Aviation to issue him with a temporary license to enable him to participate. This was granted; provided the Altair passed some rigorous testing and its owner paid a hefty Customs bond.

Finally a registration was issued (VH-USB) and while he waited for his U.S. Certificate Smithy made several record-breaking inter-city flights, partnered by P.G. Taylor. It should be remembered that at the time there was an embargo on American civil aircraft and Australian authorities were determined that their airlines used only British machines, despite the fact they did not bear comparison with a new generation of all-metal American machines coming into service.

Time was running out if he and Taylor were to be in England for the October start. When the report came through from Lockheed's it declared that the Altair was acceptable with its extra tankage. Finally the Royal Aero Club issued him with a restricted license to fly to England and start in the race, although their chances of winning were remote with the fuel restriction imposed.

Their flight to England became a desperate race in itself but any hopes were dashed following an overnight stop at Cloncurry in western Queensland. The engine cowling was found to be seriously cracked; so badly they were forced to limp back to Sydney for repairs. Damage to the cowling was more serious than anticipated and as a result there was no possibility of reaching England in time to start the race. Reluctantly Smithy informed the race committee that he was withdrawing. His efforts to start had all been in vain

It was then he was subjected to the anonymous hate mail; some even stooped to send a white feather, the ultimate insult for cowardice. Smithy was so incensed by public reaction that he responded in typical fashion. He announced that he and P.G. Taylor would make a flight to America; a much more demanding exercise than the air race. The critical stage was between Fiji and Hawaii, a distance of 3,150 miles and beyond the Altair's range, even with the additional tankage. In a clandestine operation they installed extra tanks, taking the capacity to over 500 gallons, 200 more than they were officially permitted to carry.

At 4am October 20 they took off from Brisbane en route for Fiji, the first leg in a demanding flight. Twelve hours later they put down at Suva. Their Pacific flight had none of the urgency of a race, which saw them leave for Hawaii on 29 October. Halfway to their goal they encountered an area renowned for its spectacular storms, the inter-tropical convergence zone. To add to their problems the Altair began to lose power, and as the speed fell away it entered into a terrifying spin. All of Smithy's skill went into retrieving the situation, and then to his great relief he discovered the cause of the problem. During the dramas of the storm he had inadvertently lowered the flaps. This had the effect of applying a handbrake to their progress and reduced their airspeed to a dangerous level.

At last they broke free of the storms and emerged into a clear moonlit night, and at 8.40am they landed at Wheeler field near Honolulu. Theirs had been a drama-filled flight but Taylor's navigation had been impeccable, although any euphoria at covering the distance with fuel to spare was dealt a sobering blow when on the next day Smithy took the city's mayor for a spin. Shortly after take-off the engine stopped dead, and again all of Smithy's skills were needed when he pulled off an emergency landing and the cause was soon revealed; the fuel tanks were bone dry. To their dismay one of the tanks also revealed a large split. They had reached Hawaii on a wing and a prayer

Further inspection revealed more leaking fuel tanks and a serious leak in the oil tank. Repairs took up to a week and finally they lifted off for San Francisco. Unlike the Fiji stage there were no dramas and following a 15-hour flight they landed at San Francisco. They had accomplished the first flight from Australia to The United States and in a flying time 52 hours. This was regarded by many as Smithy's greatest achievement; one that brought great acclaim but with no financial reward.

Around that time Charles Ulm and his companions disappeared on the first stage of a survey flight to Australia. Smithy was deeply grieved by the loss. His first reaction was to involve himself in an aerial search for the lost aircraft, which would have been a futile and risky exercise. Smithy's physical health was becoming a source of concern; all those years of battling the elements in unforgiving aircraft had taken their toll. His dream of opening a trans-Tasman airline was still no closer, and the Australian Government steadfastly refused him permission to use the Altair in a commercial role.

He made the decision to leave the Altair with Lockheed in the hope of selling it and returned by boat to Australia. At New Zealand he made a stopover in order to make a further appeal to the Prime Minister concerning the Tasman air service and again it was negative. Despite his failing health he began planning an attempt on the England-Australia record, currently held by Scott and Black in the race-winning Comet in a time of 70 hours and 54 minutes.

Smithy shipped the Altair to England and overcame the registration problem by adopting a British register for it (G-ADUS), but only on the condition that he not use the illegally fitted tanks. Reluctantly this was accepted; for there really was no choice. His co-pilot was Sydneysider Tom Pethybridge, a competent mechanic from the ANA days and also a reliable pilot. They took off from Croydon on 23 October 1935, and on reaching Marseilles they made a rapid fuel stop to enable them to reach Baghdad non-stop. Their hopes were dashed when they encountered a violent hailstorm, which so damaged the Altair's wing they were forced to limp back to England for repairs.

Their flight had reached desperation point for an unwell Smithy. Once more he was in deep financial trouble with his sponsors; and with insufficient funds to ship the Altair back to Sydney the record attempt was the only alternative. For the second time they left, this time from Hamble airport near London. Good progress was maintained to Greece but at Allahabad they trailed Scott and Black's time by four hours. Smithy left there; determined to make up the lost time over the next stages. The last ground sighting of 'Lady Southern Cross' was at 1.30am local time as it passed over Calcutta's Dum Dum airport en route to Singapore.

Following their non-arrival at Singapore an air search was organized, using RAF aircraft based there. Australian pilot Jim Melrose in a Leopard Moth carried out additional flights. He was actually overtaken in the darkness by the Altair on the Singapore leg and interrupted his own record attempt to conduct a personal search. Ironically he was also posted as missing but was located four days later after a forced landing on a remote beach. Finally after seven days of concentrated effort and with no sightings the search for Smithy was discontinued.

Eighteen months later some wreckage was washed up on Aye Island off the Burmese coast. This proved to be a landing wheel still attached to its oleo leg and wing framework and was positively identified as belonging to the Altair. Scientific examination of the attached marine growth established that the wheel had spent considerable time in the water.

A variety of reasons for the crash was suggested; one that the Altair had suffered sudden and terminal engine problems, possibly in the supercharger of the Wasp motor. Such was the reputation of the Wasp's reliability that this theory was largely dismissed. A more likely scenario was that a physically exhausted Smithy collapsed over the controls. Despite the Altair's rudimentary dual-control it would have been virtually impossible for Pethybridge to retrieve the situation,

An inevitable crash into the sea would have caused total disintegration of the largely-plywood Altair and no possible hope for the crew's survival. Coastal waters of the area are subject to strong tidal influences, which made it extremely difficult to pinpoint the exact location of the crash site. Precise details will never be known but with Smithy's death the world lost its greatest aviator.

**Lady Southern Cross October 1935**

# Unnamed and unloved-The DH86

DH86 'Loina'

# Unnamed and unloved-The DH86

An airliner with a dismal Australian record was the unnamed De Havilland 86, which despite its biplane configuration and fixed undercarriage was a well-proportioned and graceful design. Constructed in plywood and fabric in traditional De Havilland manner, it was powered by four Gypsy Six engines, which provided a cruising speed of 145 mph carrying 10-12 passengers. However the type did not compare with the new breed of American airliners, typified by the Douglas DC2 and Boeing 247 with their all-metal construction, retractable undercarriages, radio and other modern installations.

Qantas and Imperial Airways had forged an agreement in 1934 whereby the Imperial route from England to Australia would be shared between the two operators, with Qantas handling outgoing passenger and mail to Singapore, from where Imperial would deliver those passengers and mail to the United Kingdom and vice versa with the passengers bound for Australia. Imperial was committed to operating the DH86, and as a result Qantas was obliged to follow suit; ordering five of the type for operation on the Australia-Singapore section of the proposed Empire route. The prototype featured a single-pilot layout, which was rejected by Qantas and consequently a revised version, the DH86a was produced with dual control and in that format two aircraft were ferried to Australia and three dispatched by sea.

Holyman's Airways were the first Australian operators of the DH86 on their daily Launceston to Melbourne service across Bass Strait. Their service was inaugurated on 3 October 1934 with 'Miss Hobart', a single-pilot version; with new standards of comfort and four-engine safety being promised after the noisy and draughty Avro tri-motors which they superseded. However tragedy struck on 19 October when 'Miss Hobart' disappeared without trace en route to Melbourne.

An intensive air search was launched and continued over the following three days. Some floating wreckage was sighted in heavy seas off Wilson's Promontory but could not be located by vessels involved in the search. An official enquiry offered a possible explanation that the accident occurred during a change-over of pilots. If such was the case it was a dangerous procedure in the cramped conditions of the single-pilot DH86. The aircraft would have been flying at low altitude, possibly less than 1,000 feet and with little chance of recovery if it went out of control. Structural failure was considered unlikely in such a new aircraft.

However the DH86 saga had only just begun; when on 15 November there came the startling news that Qantas's VH-USG was destroyed in a fatal crash. This latest tragedy happened after an early morning take-off from Longreach in western Queensland; with the aircraft in the final stage of its delivery flight from England to Brisbane. On board was a crew of three and one passenger who joined the flight at Darwin. Rescuers at the crash site found the aircraft's fuselage completely collapsed; three of the occupants were deceased and a third close to death. A mystifying discovery was the flight engineer occupying the captain's seat; while the captain's body was located at the rear of the aircraft. Eye witnesses described how the aircraft was sighted flying normally at 1000 feet when it turned to starboard and entered a spin from which it never recovered.

Possibly the captain was taken unwell, vacated his seat and handed over to the flight engineer. VH-USG was carrying an amount of spares plus a complete engine that was lashed down in the centre section of the aircraft. One theory proposed was that the weight of the captain, a big man, had upset the balance. It was later judged that the DH86 was unforgiving in a crisis and beyond the abilities of the flight engineer. Crash investigators also expressed concern at apparent defects in the forward fin post, posing the question that a failure in that component had caused the loss of 'Miss Hobart' and VH-USG.

Qantas were sufficiently concerned to question De Havilland's workmanship, and at one stage they contemplated making legal claims against them. A further dilemma was the widely advertised Brisbane- Singapore Air Mail, scheduled to commence on 10 December 1934. With the airworthiness of the DH86 in question, Qantas reverted to using two elderly DH51s to inaugurate the service. Not surprisingly De Havillands were outraged by Qantas's inference of bad workmanship. Meanwhile a second DH86 (VH-USD) was found to have the fin bias mechanism cracked in the manner of the lost VH-USG. Argument raged back and forth; with De Havillands unprepared to admit to any design fault.

Qantas carried out its own modification to the suspect fin mechanism and with Civil Aviation Branch approval the DH86 entered Qantas service. As a precaution no fare-paying passengers, only mail was carried for a period of three months.

In early 1935 a second DH86 was accepted by Holyman's Airways. During its assembly before entering passenger service Holyman's paid great attention to the suspect fin area. Their precaution was justified when it revealed an incorrect assembly of the unit. Qantas's earlier inference of faulty workmanship would seem to be vindicated.

Meantime, Holyman's new acquisition 'Loina' (VH-URT) entered service and was joined three months later by 'Lepena' (VH-USW). In late September a third DH86, 'Loila' made up the trio. On 2 October 1936 'Loina' departed Melbourne's Essendon aerodrome on a scheduled flight to Launceston; on board were three passengers and two flight crew. Shortly afterwards 'Loila' left Melbourne on its proving flight to Sydney; carrying six passengers; these were a mix of trade reps. and journalists.

After a leisurely flight 'Loila' touched down at Sydney's Mascot aerodrome, but any euphoria of the occasion was soured by the startling news that 'Loina' had vanished. Its last sighting was at 9.30am off Wilson's Promontory, with a final radio call twenty minutes later, prior to set-down at Flinders Island. With no response to Essendon's radio calls and its non-arrival at Launceston the alarm was raised. 'Lepena' began an immediate search of the area and their worst fears were confirmed with the sighting of wreckage off Flinders Island.

Over the following days an amount of wreckage was retrieved for the investigating team, yet such was the fragmentation of the aircraft it was impossible to pinpoint a definite cause to the crash. Qantas and Holyman's persevered with their DH86 fleet until they were replaced during 1938 with the Empire flying boats and Douglas DC2s respectively. New operators were found for the surviving DH86s; including the RAAF who used some as trainers for wireless operators. One example served as an air ambulance in the Middle East campaign. This aircraft had the misfortune to be intercepted by Luftwaffe Me109s, and despite its Red Cross markings was ruthlessly shot down.

The rogue airliner made a final statement on 20 February 1942 when VH-USE crashed to destruction shortly after leaving Brisbane's Archerfield aerodrome. A crash investigator arrived on the scene only to find that the wreckage had been deliberately burnt, thus destroying any opportunity to solve the airworthiness enigma. Significantly the aircraft's fin was discovered intact almost a mile from the crash site, showing clearly that it had detached itself in the air. Perhaps that troublesome unit was determined to have the last word in a chain of tragedies. By coincidence this was the very day that Darwin was bombed for the first time and consequently the loss of an elderly airliner was barely newsworthy by comparison to the Darwin experience.

# The missing Stinson 1937

## Bernard O'Reilly

## The Missing Stinson

In March 1936 the 'A' series Stinson tri-motor appeared on the Australian scene. Three aircraft were purchased by Airlines of Australia, a new company on the local scene which offered a daily service from Brisbane to Sydney and Sydney to Melbourne. Also available was a service from Brisbane to Townsville, operating four times a week.

With comfortable seating for eight passengers and a high cruising speed they offered a great advance over existing types such as the ageing Avro Tens. A fourth aircraft joined the fleet later in the year; and during that period the company built up an enviable record for reliability and safety. They became a familiar sight along Australia's seaboard with their distinctive blue and orange livery and that evocative whine from their American Lycoming engines.

Friday 19 February 1937 was an overcast and windy day when VH-UHH, 'City of Brisbane' left Brisbane's Archerfield aerodrome at 1.00pm on a scheduled flight to Sydney. Five passengers boarded the aircraft, which was scheduled to stop en route at Lismore to embark several more. Weather forecast for the northern rivers of New South Wales was for heavy cloud and widespread rain showers. Such conditions were a feature of those sub-tropical areas and presented no problems for the experienced flight crew. Captain Rex Boyden, a pilot with wartime service in the Royal Naval Air Service was chief pilot.

Captain Beverly Shepherd was not actually rostered for duty but was taking the opportunity to travel to Sydney. This was to fulfil a social appointment with Jean Batten, the New Zealand aviatrix who was temporarily staying in Australia and was reported to becoming engaged to Shepherd.

Lismore had notified Brisbane of conditions there; with heavy cloud in the area and the airfield affected by water. 'Brisbane' was scheduled to land at Lismore at 2pm but with their non-arrival it was presumed they intended to fly non-stop to Sydney. By late afternoon there was no word of 'Brisbane', which led to a sense of alarm when it failed to arrive in Sydney. Significantly the aircraft carried no radio, making it possible that it had force-landed in an area without access to a telephone.

Darkness fell, and with still no news the company faced the awful possibility that it had crashed. At first light, four of the company aircraft began a search of the Stinson's route. They were joined later by units from the RAAF and a variety of civil aircraft in what developed into a most intensive search. Witnesses who claimed to have seen or heard the aircraft in the Broken Bay region north of Sydney reported many sightings. Consequently that area became the main focus of the search. Days passed with no discovery until it was abandoned; with the conclusion was that the Stinson had come down in the sea off Broken Bay.

Despite claims of those who believed the Stinson had crashed north of Sydney, there was one person who had his own theories about the missing aircraft. Bernard O'Reilly was convinced it would be found in the rugged MacPherson ranges. He was an experienced bushman who lived in the remote rainforest area of the Lamington Plateau. Despite the fact that eight days had elapsed since its disappearance he made the decision to set out along the intended flight path. For one full day he trekked up and down gorges and through jungle-like forest where few humans had ever ventured. At noon on the second day he caught sight of a burnt-out section on a far-off ridge that could only have been caused by a recent blaze such as an aircraft crashing and burning.

By late afternoon he reached the site, and in a mix of emotions he discovered the burnt-out Stinson wedged at the base of two trees. Incredibly there were two survivors; with one in a pitiable condition lying near the wreck. John Proud had suffered a fractured leg that was badly infected. Joseph Binstead was in better shape and had been caring for Proud as best he could. That involved a painful daily trek to a stream to fetch water; apart from that they had existed on berries from nearby trees.

A third person, James Westray had survived the crash and although inexperienced in survival in the Australian bush he set out down the valley to find help for Proud and Binstead. Both pilots and two other passengers had perished in the crash and ensuing fire. The survivors described the terrifying storm and its violent winds that literally forced the aircraft down into the rain forest. The initial impact was softened as they crashed through the tops of the taller trees until two large ones halted their progress. At that point the Stinson slid downwards to their base and only for the fire all the passengers may have escaped. They were of the opinion that both pilots died on impact.

O'Reilly made the pair as comfortable as possible and then set off westwards to find a farm or settlement and organize a rescue party. On his way he came across Westray's body beside a stream. He had suffered a broken ankle, and being unable to move he had perished there.

Night had fallen before he reached civilization; where he was alerted to the sound of a rifle. It transpired that it was a settler shooting flying foxes in the dusk; and was able to provide two horses for them to continue the rescue. They rode through the night to find the nearest property with a telephone; and from there the news of the discovery was relayed to an amazed nation. Despite his overwhelming fatigue O'Reilly led a rescue party that included a doctor back up the gorge to the crash site.

It was late morning before the exhausted team reached the area, and fortunately the doctor was able to save Proud's damaged leg. Hot food and warm clothing was provided for the survivors to make them comfortable for the first time in eleven days. They were then stretchered back to civilization by teams of bearers through the rain forest to waiting ambulances, which was an epic in itself. Both men made a complete recovery, thanks to the unstinting efforts of Bernard O'Reilly, who later wrote an account of the mission in an autobiography titled *Green Mountains.*

Bernard O'Reilly died in 1975 but the memory of his epic rescue is well documented at the O'Reilly Guest House on Lamington Plateau. Apart from a plethora of memorabilia at the gift shop there is a magnificent life-size set of bronze figures depicting the moment of rescue. Today the crash site is a source of interest for bush walkers who can pay homage at the graves of the unfortunate victims and to view the skeletal remains of the Stinson that are still embedded in the trees that claimed it on that stormy day in 1937.

**City of Brisbane May 1937**

# First Australian World War 2 Ace (1940)

**Lesley Clisby**

# Les Clisby

Australian fighter pilots featured strongly from the onset of World War 2 and during its six year duration many achieved 'ace' status, although the RAF showed a preference not to acknowledge such a system. The first to attain this was Lesley Redford Clisby, as a member of the renowned No.1 Squadron based in France during the brief, hectic period of the German Blitzkreig, which was unleashed on 10 May 1940.

Clisby's eventual involvement with the RAF's long-established No.1 squadron began in 1935, with his enlistment as ground crew in the RAAF at Point Cook, Victoria. Born at McLaren Vale South Australia in June 1914, Clisby showed an aptitude for mechanical engineering and after a term as ground crew he applied for an officer cadet course in 1936. During flight instruction he was forced to parachute from his Gypsy Moth, which was on the verge of crashing. Surviving that episode, Clisby graduated at Point Cook in June 1937, and applied for a short term commission in the RAF.

After further instruction at No.1 Flying Training School at Leuchars in Scotland, he was posted to the leading fighter squadron of the day (No.1 Sqdrn. RAF). This unit had only recently exchanged its agile Hawker Fury biplanes for the purposeful 8-gun Hawker Hurricane, and thus equipped the squadron was posted to France, landing at Le Havre on 8 September 1939. Four Hurricane squadrons were deployed to France that September; Nos. 1, 73, 85 and 87, forming the fighter element of the Air Component, whose task was to support the BEF and provide where possible fighter escort for the Blenheim and Battle bomber squadrons also based in France.

The ensuing six months was a period remembered as the 'Phoney War'; a time when the squadron flew many patrols and was moving from airfield to airfield, until finding a permanent base at Vassincourt. During this period the value of the Hurricane's sturdy wide-track undercarriage was much appreciated by pilots and ground crew alike. French airfields in the main lacked adequate drainage and the soft ground would have wrought havoc with a weaker landing gear.

Early in the New Year No.1 squadron received a fresh complement of Hurricanes, these examples powered by the up-rated Merlin III, driving variable pitch airscrews, replacing the earlier two-bladed wooden propellers, which proved inadequate in operational use.

The winter of 1939/40 was a particularly severe one; and consequently little action from the Luftwaffe taking place; however in March 1940 there was more intense activity. May 10 saw the launch of the Blitzkreig, a time when the Luftwaffe came in force to support their ground attack. On 12 May Clisby was credited with the destruction of six enemy aircraft (in that one day), an exploit which earned him the DFC.

The following days saw even more intense enemy attacks, heavier; it was later recorded than those that occurred during the Battle of Britain. Clisby's brief but hectic combat career ended 14 May, following an engagement with a formation of Me110s. Clisby and his good friend 'Lorry' Lorimer both failed to return from this mission. Clisby's victory tally was thought to be twenty, of which 14 had been destroyed in the three days before his death. Post-war research found that the French had discovered Clisby's body in the burnt-out remains of his aircraft and buried him in a temporary grave. Later the War Graves Commission re-buried him in the Commonwealth War Graves cemetery at Chuloy, near Nancy, France. He is also remembered on the War Memorial in Adelaide and on the Roll of Honour at the Australian War Memorial at Canberra.

**Hawker Hurricane/Heinkel III combat (May 1940)**

# First W/W2 Australian air VC (1941)

**Hughie Edwards V.C. DSO DFC**

# Hughie Edwards VC

With the fall of France in July 1940, Western Europe, apart from Spain and Portugal fell under Nazi subjugation. Great Britain finally stood alone to face the anticipated Battle of Britain, to be followed by 'Operation Sea Lion', the projected seaborne invasion of the British Isles. Although vastly inferior in numbers to a rampant Luftwaffe, RAF Fighter Command would emerge triumphant from that dramatic encounter, arguably the most decisive battle of the twentieth century. Throughout that turbulent, early phase of World War 2, RAF Bomber Command offered the one opportunity to carry the offensive into the Nazi aggressor's domain. Despite the daunting casualty rate, Australia provided Bomber Command with some notable airmen; not the least of these was Hughie Idwal Edwards.

The future Air Commodore and VC winner was born in Western Australia in 1914 of a Welsh immigrant family. His determination to pursue an aviation career was realised in 1936 with his acceptance as a cadet in the Royal Australian Air Force. To further his career, in 1938 he transferred to the Royal Air Force, where he continued his training on the newly-introduced Bristol Blenheim. In the course of a training flight Edwards was forced to make an emergency bailout, and had the misfortune to strike the aircraft's tailplane and suffer severe damage to his leg. That situation was further aggravated in the ensuing heavy parachute landing.

So severe was the problem that his flying career appeared to be finished. However, by dint of determination and persistence with medical boards he finally achieved a full flying category. During a night-flying exercise in 1940 Edwards suffered a further setback; when a Luftwaffe raid in his sector forced the closure of local airfields and total radio silence. He elected to remain with his aircraft and face the daunting prospect of a forced landing in a blacked-out countryside. In a collision with a tree Edwards suffered severe concussion, a factor that prevented him from becoming operational until early 1941. His posting to 105 squadron in Norfolk coincided with a period of maximum effort of daylight raids on shipping and heavily defended objectives in occupied Europe. At that stage of the war the Blenheims of No.2 Group were approaching obsolescence and suffering heavy operational losses, nevertheless operations continued unabated.

Edwards was awarded the DFC during that period and attained the rank of acting wing commander, due to losses and his own fortuitous survival. Germany had invaded Russia in June 1941 and to draw elements of the Luftwaffe from the Eastern Front the British War Council began a policy of heavier attacks on German-occupied territory.

On July 4, units from No.2 Group were ordered to carry out a significant raid on the port of Bremen in northern Germany. This followed a night attack from Bomber Command on the previous evening. Edwards was chosen to lead the risky daylight mission; comprising a combined force of nine Blenheims from 105 squadron and six from 107. He led the formation at wave level over the North Sea and as they neared the target he made a turn to the northeast. His plan was to approach Bremen from a northerly direction in the hope they might achieve an element of surprise.

However, Bremen's defences were fully alerted as the Blenheims swept in, lined out in single file; with each pilot selecting an individual target. The formation ran the gauntlet of barrage balloons and vicious flak, heavy enough to account for two aircraft that were brought down over the target while another turned inland and failed to link up with the formation. Considerable damage was inflicted on the dock area and warehouses, after which Edwards skilfully marshalled his forces and led them home with no further loss

A fortnight later; following the Bremen raid Edwards was awarded the Victoria Cross, the first to an Australian airman in World War 2, the citation of which concluded:

*Throughout the execution of this operation, which he had planned personally with full knowledge of the risks entailed, Wing Commander Edwards displayed the highest possible standard of gallantry and determination.*

In a gesture of appreciation Edwards framed the citation and presented it to the squadron, maintaining that the award was a team effort and not an individual one. Shortly afterwards the squadron was posted to a beleaguered Malta where it operated on shipping strikes and once again Edwards survived his tour despite the inevitable heavy losses.

On his return to the U.K. he was selected to accompany a group of RAF commanders on a goodwill tour of The United States and Canada. In December he was operational again with 105, which had replaced its obsolete Blenheims with De Havilland Mosquitoes; the first RAF squadron to operate with this exceptional aircraft. On 6 December 1942 he led a combined force of Mosquito, Boston and Ventura bombers on Operation Oyster, a large-scale daylight raid on the Philips electrical factory at Eindhoven in Holland. RAF losses were heavy, with 14 aircraft brought down by flak and fighters. Damage to the factory was substantial, while few casualties were suffered by Dutch workers.

Edward's leadership on this and other missions was recognised with the award of the DSO. Following this was his tenure as Station Commander at RAF, Binbrook, and home base for the celebrated No 460 Lancaster squadron (RAAF) of No.1 Group. Despite the administrative duties of the command Edwards carried out a number night operations himself.

In December 1944 he was promoted to Group Captain and posted to South East Asia Command where he occupied various senior posts. He remained in the Command until 1947, returning to the United Kingdom to attend the RAF Staff College. Edwards returned to flying duties in 1950 and between 1953 and 1956 commanded a jet fighter base before being posted to Iraq for a three- year term as station commander of the RAF Base at Habbaniyah.

Edwards was accorded further promotions and honours; notably as Air Commodore and Commandant of the Flying Wing at RAF Brize Norton until his retirement from the Royal Air Force in September 1963. Edwards returned to Western Australia and began a successful business career. In 1974 he was knighted and appointed Governor of Western Australia. It was a position that he was forced to resign the following year due to ill health.

In company with former squadron companion and Test cricketer Keith Miller, Edwards was about to attend a match at Sydney Cricket Ground in 1982 when he unexpectedly collapsed and died, aged 68. Hughie Edwards was a remarkable warrior, the most highly decorated Australian of World War 2 and one of the giants of Bomber Command; deservedly remembered alongside Guy Gibson, Leonard Cheshire, Donald Bennett and a host of other luminaries.

**Bremen Raid (4-7-41)**

# The Outback Aussie

**R.H. Middleton V.C.**

# Middleton VC

Rawdon Hume Middleton (Ron to his mates) was born in Sydney in 1916. He completed his secondary education at Fort Street Boy's High, but an academic career held no appeal for the young man; instead he spent his teen years as a 'jackaroo' at an outback New South Wales sheep station. Middleton was an introverted and reticent teenager; subject to moods of melancholy and with a strong preference for his own company; in many ways he was the typical outback Aussie.

Anxious to achieve aircrew status in World War 2 he joined the RAAF in 1940 and was accepted for the pilot's course. His draft sailed for Canada in 1941 where they continued training. He was the earnest, plodding type of pupil, steady and thorough if not spectacular. Following that period Middleton and his fellow airmen sailed to the United Kingdom in 1942. He was posted to No.149 heavy bomber squadron in Suffolk and from there he began operations as second pilot on Short Stirlings. In April 1942 he was second pilot on a raid on Essen. Over the target they were coned by searchlights and then came under attack by a night fighter. Their Stirling was subjected to repeated assaults, sustaining damage to the starboard wing and setting fire to an engine. They managed to escape the situation and limp back to base, where the aircraft broke up completely on landing. Most crews would be unnerved by the experience, but in Middelton's case it was the complete reverse. It roused him from his melancholy and galvanised him into a companiable mood which never left him. Perhaps the Essen experience had banished any self-doubts about his courage?

In July he was promoted to captain and over the ensuing months Middleton and his crew worked their way steadily through an operational tour. Targets included Hamburg, Munich, the Ruhr and other German targets. Following these operations there was a succession of raids on northern Italy. On the night of 28 November they faced another Alps crossing. Their target on this occasion was the Fiat works at Turin and the mission was their twenty ninth; the penultimate before the completion of a tour. Raids on Italian targets were treated with a degree of levity; with those missions being displayed on aircraft as ice cream cones rather than the traditional bombs. Their greatest peril was the Stirling's inability to achieve its designated altitude when fully loaded. On occasions an aircraft was forced to fly between the unforgiving Italian Alps rather than over them; many an aircraft that failed to return from operations had crashed to destruction in these hostile mountains.

Middleton's Stirling (H for 'Harry') was one of seven selected crews from 149 squadron who made up the 180-strong force dispatched. The outgoing journey developed into a nerve-wracking experience as they picked a hazardous course between the towering peaks. Middleton was about to jettison the bombs and abort the mission, when flares from the Pathfinder Force appeared in the distance over Turin. They had made it on a wing and a prayer but were then faced with a barrage of flak as they descended towards the city centre and their target. Middleton flew a cautious path between the tangle of parachute flares and the flak that was increasing in ferocity. At that moment they sustained a hit on their port wing, causing the aircraft to stagger off course. Middleton regained control and made a wide circuit before he began a second run up to the target; and then with barely half a mile to cover the Stirling received a direct hit to the windscreen.

Both pilots were concussed and Middleton collapsed over the controls; grievously wounded. In spite of his own wounds Hyder the co-pilot dragged Middleton clear and took control. Middleton regained consciousness; and despite the loss of his right eye and other serious wounds he then insisted on making a third attack which was completed successfully. They were now faced with a four-hour flight across Occupied Europe in a crippled aircraft with its dying captain.

Despite being almost blind and barely able to speak, Middleton remained at the controls throughout the flight. On arrival at the English coast in the pre-dawn he turned, flew a parallel course and issued orders for the crew to bail out. Five members made successful descents while two remained to assist Middleton. He headed seawards to avoid the possibility of crashing into houses and as he prepared to ditch; at that critical moment the bomber's engines died. The remaining crewmembers bailed out but with insufficient altitude they perished in the attempt. With its fuel tanks empty the Stirling crashed into the sea; taking its gallant captain with it.

In February 1943 Middleton's body was washed ashore at Dover. A fortnight earlier his devotion to duty was recognized with the award of a posthumous Victoria Cross. The citation concluded thus:

*While all the crew displayed heroism of a high order, the urge to do so came from Flight Sergeant Middleton whose fortitude and strength of will made possible the completion of the mission. His devotion to duty in the face of overwhelming odds is unsurpassed in the annals of the Royal Air Force*

**Turin mission (28/29-11-42)**

# Sportsman and Fighter Ace

**Keith Truscott**

# Keith (Bluey) Truscott

Future fighter ace Keith William Truscott was born 17 May 1916 at Prahan, Melbourne and proved to be good scholar, while as a teenager excelled at sports, going on to be a key member of the Melbourne Football Club from 1937 to 1940. His auburn hair inspired the nick-name 'Bluey', which is an inevitable outcome for an Australian blessed with reddish hair. His decision to enlist in the RAAF attracted much publicity, and surprisingly for someone of his mental and physical abilities he struggled with flying lessons. The fact that he was a prominent sportsman was in his favour, being granted extra time in which to qualify.

In September 1940 Truscott was granted leave to play in Melbourne's winning Grand Final side and in November that year he arrived in Canada under the Empire Air Training scheme. Commissioned in February 1941 Truscott was posted to the newly-raised 452 squadron RAAF, based at Kenley Lincolnshire, from where it carried out fighter sweeps and bomber escort missions, operating with the improved Spitfire IIa.

Truscott's original flight commander was the celebrated Brendan 'Paddy' Finucane, who was destined not to survive the year, being lost over the English Channel, following a mission over France. During Truscott's tenure with 452 he was credited with eleven confirmed victories, was awarded the DFC and Bar and by October 1941 became the squadron's C.O.

In May 1942, with Australia facing the possibility of a Japanese invasion the squadron was returned home, and based at Darwin was charged with repelling the Japanese air raids that were increasing in intensity. Truscott however was posted to No.76 squadron, a Kittyhawk fighter-bomber unit recently moved to south-east Papua. This was in July 1942, only weeks before the anticipated Japanese landings at Milne Bay to their east.

The two Kittyhawk squadrons, 75 and 76 played a decisive role in crushing the assault, which was memorable as being the first time in World War II that a Japanese land offensive was defeated. At a critical stage of the campaign, the conditions were quite appalling, with near-constant rain, mist, low cloud and a perilously slippery air strip, During these adverse conditions No. 76 squadron's commanding officer was killed. Truscott assumed command; leading from the front in his dynamic style until the Milne Bay action was complete, a time when he was mentioned in despatches for his involvement.

Following the battle of Milne Bay, 76 squadron was transferred to mainly tedious garrison duties in north-west Australia, although Truscott did manage to bring down a Japanese bomber, increasing his tally to fourteen enemy aircraft destroyed, three probables and three damaged. Sadly time was running out for 'Bluey' Truscott, when on 28 March 1943, while carrying out mock attacks on a Catalina flying boat over Exmouth Gulf he misjudged the height, struck the water and was killed. He was buried with full air force honours at Karrakatta cemetery, Perth.

Spitfire Mk.II

# Salamaua VC (1943)

**Bill Newton VC**

## Newton VC

William Ellis Newton was one of three Australian airmen to win the Victoria Cross in World War 2. He was born in Victoria in 1918 and joined the RAAF in 1940. On gaining his wings he served as an instructor until May 1941, and was then posted to 22 Squadron, an operational unit, serving in the South West Pacific area. This was the only Australian unit to operate with the Douglas Boston; a type that served effectively in every war theatre.

Flight Lieutenant Newton established himself as a determined combat pilot during that period; carrying out 50 operational missions. On 16 March 1943 he was involved in a hazardous daylight raid on fuel and storage dumps at Salamaua. Newton's aircraft was hit by anti-aircraft fire that took out one engine and caused damage to the airframe and hydraulics. Despite these distractions he successfully attacked the target and limped back to base.

Two days later on what was his 52$^{nd}$ mission he returned to the area to complete the destruction of a previously damaged building. This was the focus of the raid and was heavily defended with anti-aircraft guns. During Newton's run up to the target his aircraft suffered fatal damage. He still pressed home his attack and scored a direct hit on the building. The crippled Boston staggered from the scene, and with one engine in flames Newton endeavoured to clear the area and save his crew but such was the damage he was forced to make an emergency ditching.

Newton swam ashore in company with his observer Sergeant Lyons; the third crewmember did not survive the crash. The two airmen evaded Japanese patrols for only one day before they were captured, and shortly afterwards were transferred to Lae for interrogation. One can only speculate on the brutality of this episode, for they were then returned to Salamaua, and on 29 March both airmen were summarily executed. Newton suffered death by beheading which was considered honourable according to the Samurai code. In reality it was an atrocity and contrary to the rules of war. Sergeant Lyons was tied to a tree and bayoneted to death. His identity was not revealed until October 1948. Following the capture of Salamaua in October 1943, Newton's fate was established after identification of his body.

For his exploits on the March 16 raid Newton was awarded a posthumous V.C. The official citation of the award concluded thus: *Flight Lieutenant Newton's many examples of conspicuous bravery have rarely been equalled and will serve as a shining example to all who follow him.*

Salamaua action 18 March 1943

# Top-scoring RAAF Ace

**Clive Caldwell**

# Clive Caldwell

Clive Robertson Caldwell gained fame as the top-scoring Australian fighter pilot of World War 2. Tall and athletic, he was an excellent marksman, a brilliant pilot and potential leader of men. In a hectic career that spanned four years of combat he flew Tomahawks and Kittyhawks in North Africa and Spitfires in Europe and the South West Pacific. During that period he was in the unique position of having fought the Germans, Italians, Vichy French and the Japanese and shot down at least 30 enemy aircraft. The future ace was born in Sydney in 1910, a time when the aeroplanes that were to play a large part in Caldwell's career scarcely existed. It is worth noting that twelve days prior to his birth a young Australian, John Duigan had built and flown the first locally-designed aircraft at Spring Plains in Victoria. Three years would elapse before Australia set up its own air force with the formation of the Australian Flying Corps at Point Cook in Victoria in 1913.

Caldwell was born into a middle-class Sydney family and was educated at Trinity Grammar where his academic career was generally unremarkable, although he excelled in athletic pursuits and boxing as a teenager. His first employment situation was with the Bank of New South Wales but the restless Caldwell rebelled against the rigidity of the banking system and found other employment in a variety of jobs. One of these was as a jackaroo at an outback sheep station where in off-duty periods he was able to hone his shooting skills, to the detriment of a future generation of opposing airmen.

In 1938 Caldwell's finances were sufficient for him to take flying lessons at the Royal Aero Club of N.S.W. After only three and a half hours of dual instruction he made his first solo and with the approach of war he had accumulated time on Tiger Moths. With the outbreak of war in September 1939 Caldwell was eager to be a fighter pilot, however the cut-off point for single-seater training was 28 and only by altering the details on his birth certificate did he manage to be accepted into the RAAF. To his dismay he discovered that those on his course were to be trained as instructors, a situation which held no appeal to Caldwell. He promptly discharged himself and in April 1940 for a second time he joined the RAAF, this time as a trainee in the Empire Air Training Scheme. His competence as a pilot was obvious and on gaining his wings he was impatient to be sent to England to fight in the Battle of Britain, however by October it was all but over. Instead he embarked for the Middle East in February 1941, where, as Pilot Officer Caldwell he joined 250 squadron RAF based at Aquir in Palestine.

This unit operated with the Curtiss P40C Tomahawk which had performed valiantly as a front-line fighter with the US Army Air Corps during 1941 and 1942, although in RAF service it was rejected as a fighter in Western Europe. Their Allison engines were rated to only 12,000 feet, rendering combat above 15,000 feet totally impractical; fortunately for the P40 pilots in their encounters with the BF109 and Macchi 200 the desert war was generally a low-level affair. Initially, Caldwell like so many other tyro pilots had little success in his early sorties, until his 33$^{rd}$ mission when he scored his first confirmed kill; a BF109 plus a share in the destruction of a Cant Z1007 bomber near Alexandria. Caldwell's reputation as an air fighter became legendary, as witnessed by the mounting number of crosses stencilled on his P40. In January 1942, with his victory total at 18 Caldwell was promoted to squadron leader and given command of 112 squadron RAF.

This was a commendable achievement, being the first Empire Air trainee to command a squadron and with the prestige of the award of the DFC and Bar. This unit had recently converted to the improved version of the Tomahawk, the P40E Kittyhawk. Although inferior to their main opponent, the BF109 they performed well in a variety of roles, particularly in bombing and strafing of enemy transport and airfields.

Caldwell's tenure with 112 ended in May 1942 when he was ordered to report to England. At that point his victory tally was 20½ kills after 550 hours of operations. His passage to the United Kingdom was performed over a series of flights via West Africa, The Caribbean, The United States and Canada, finally arriving in Scotland in late May 1942. By June he was back on operations commanding the Kenley Spitfire Wing, and in a period of ground attacks over France he added a locomotive to his list of kills.

With Japan having entered the war, Australia was experiencing air raids on a regular basis on Darwin and other Northern Territory towns. As a result Caldwell was recalled to Australia in November 1942 as wing leader of the newly-raised No.1 Fighter Wing. This unit, which comprised former UK-based Spitfire squadrons No.452 RAAF, No.457 RAAF and No.54 RAF arrived in Darwin in January 1943. Their aircraft were Spitfire Vcs fitted with tropical filters to cope with local conditions. The fighter wing began operations over northern Australia with its focus on the defence of Darwin which at the time was a major factor in the defence of Australia.

On 2 March 1943 Caldwell claimed his first Japanese victories while leading a flight of six Spitfires. His formation intercepted 6 Kate bombers escorted by 12 Zero fighters about to attack Allied shipping in the Arafura Sea north of Darwin. In the ensuing melee Caldwell destroyed one Zero and one Kate and by August Caldwell had despatched 8 Japanese aircraft, elevating his tally 30.5. He was then taken off operations and posted to No.2 Operational Training Unit as Chief Flying Instructor.

In May 1944 Caldwell returned to operations as wing leader of No.80 Fighter Wing based in Morotai. This unit which comprised No.79, No.452 and No.457 squadrons operated with the far-superior Spitfire Mk.VIII and rather than the fighter role it was employed in ground attack sorties. In fact their Spitfires were ill-suited to these duties, proving quite vulnerable to ground fire in what the pilots regarded as unimportant targets.

By the end of 1944 it was apparent they were being left out of the main arena of operations and as a result of nothing having been done to meet the pilots' demands Caldwell and two other senior officers resigned in protest. This affair gained notoriety as the 'Morotai Mutiny' and led to a command crisis in the RAAF. An investigation led to two senior officers being relieved of their appointments. Caldwell was eventually reinstated and finished the war attached to HQ Ist Tactical Air Force RAAF. He resigned from the RAAF in 1946.

In a post-war situation Caldwell created a wholesale fabrics business. Failing health caused him to relinquish his management of the enterprise and on 5 August 1994 Australia's top-scoring ace passed away; remembered as Australia's top-scoring fighter pilot, and is listed in the top ten allied aces.

**Spitfire Mk.Vc (Darwin 1942)**

# Master Airman

**Air Vice Marshal D.C.T. Bennett**

# Air Vice Marshal Donald Bennett

To be accorded 'Master Airman' is indeed a lofty title, and among those whose careers justified such an inclusion, one person stands tall. He was Donald Bennett, superb pilot and navigator, company director and for a brief post war period a Liberal Party politician.

Donald Clifford Tindall Bennett was born at Toowoomba, Queensland in 1910, the youngest of four brothers to stock and station agent and grazier George Bennett and his English-born wife Celia. His three brothers all excelled in the Law and Medicine, while Don's current lack of academic distinction found him working as a jackaroo on his father's cattle station. His later studies as a science student, together with his militia involvement, resulted in a successful application as a cadet in the RAAF. Joining up in July 1930 he began pilot training at Point Cook, Victoria, and at the end of the course came top in practical flying and second in the theoretical examinations.

At the conclusion of the course the cadets were informed that no positions existed in the RAAF, due to the strictures of the Great Depression, however they could accept a short-term commission in the RAF, to which Bennett and several others agreed; thus launching a life-long association in the Home Counties for Pilot Officer Bennett. He served for a period with No. 29 (Fighter) squadron, a unit which operated the unpopular Siskin biplane; however he was keen to be involved with flying boats, which saw him posted to the flying boat station at Calshot, where he did particularly well at navigation.

His next posting was to No.210 squadron at Pembroke Dock in south-west Wales, which operated the large open-cockpit Supermarine Southampton biplane flying boats. By coincidence the new C.O. was Arthur Harris, who as C-in-C of Bomber Command during World War 2 was to have a profound influence on Bennett's career.

In 1935 Bennett resigned from the RAF and transferred to the RAAF Reserve. During his RAF tenure Bennett gained a remarkable array of licences; a first class civil navigator's licence, a wireless operator's licence, three categories of a ground operator's licence, a flying instructor's certificate and significantly a marriage licence, following his registry office ceremony to wed an attractive Swiss woman, Elsa (Ly) Gubler.

Still only aged 26 Bennett joined Imperial Airways and was shortly promoted captain of the large current airliners, such as the biplane Handley Page 42 on European routes. With the introduction of the 'C' class Empire flying boats, Bennett was allocated 'Cassiopeia' for operation on Imperial's expanding overseas destinations.

A revolutionary project, with which Bennett was involved, was the Short-Mayo composite, where a four-engine seaplane 'Mercury' was launched from the back of its mother flying boat 'Maia', enabling the heavily-laden Mercury to achieve long-distance flights. In July 1938 Bennett made the first commercial transatlantic crossing with Mercury, carrying newspapers and film to New York from the United Kingdom. In October he flew Mercury non-stop from Scotland to South Africa to establish a long-distance record for seaplanes, which remains unbroken to the present time.

Following the outbreak of war in 1939, during 1940 Bennett was approached by the Minister for Aircraft production, the dynamic Lord Beaverbrook with the proposal that he explore the possibility of creating an Atlantic Ferry Service. As its flying superintendent Bennett organised the inaugural ferry mission, leading the first flight of seven Lockheed Hudsons from Canada to the United Kingdom. Having seen the Atlantic Ferry to a successful beginning, Bennett re-joined the RAF in September 1941, as an acting wing commander at an elementary navigation school. In December he was given command of 77 squadron; a heavy bomber unit still operating the obsolescent twin-engine Whitley, and during his tenure he consistently flew on operations.

In April 1942 and during his term as C.O. of No. 10 squadron, a unit that operated with the four-engine Halifax, Bennett was brought down by ground fire during an attack on the battleship Tirpitz, anchored near Trondheim in Norway. He managed to bail out his crew and himself and linking up later with a crew member the pair made a remarkable overland trek to neutral Sweden. After a brief internment there, Bennett was flown back to England, and for his achievement was awarded the DSO.

Promoted to acting group captain in July 1942, Bennett was directed by Air Chief Marshal Sir Arthur Harris to form and command the Pathfinder Force within Bomber Command. Up to that time, bombing results were lamentably poor, and with a loss rate of 5% of aircraft sent on operations, Bomber Command was achieving very little and at great cost. Therefore, a force to find and mark targets for Bomber Command was deemed essential, and Bennett's appointment, with his superlative skills was to be crucial to its success. In January 1943 the

Pathfinder Force had become No.8 Group, Bomber Command, and in December at the age of 33, Bennett was promoted to acting air vice marshal, the youngest officer to ever hold such rank, and with it numerous awards.

The war had a sour ending for Bennett, for of all senior group commanders, he alone was not knighted. This could be attributed to his personality, which was described as impatient, dictatorial and pedantic, and while effective it created enemies from less efficient RAF contemporaries. Bennett was released from the RAF in May 1945, in order to contest a by-election in the House of Commons. He was elected unopposed, but was defeated at the general election in July, and his brief period as an MP ended in disillusionment and argument.

In 1945 Bennett had been elected chief executive of the emerging British South American Airways Corporation; however his policy of using only British-made equipment, principally the Avro York and Avro Tudor contributed to a series of tragic accidents, resulting in the Tudor being grounded. Bennett's criticism of the board of BSAA was not well received and as a result he was asked to resign. During the 1948 Berlin Airlift, using Tudor freighters, he formed a profitable air transport company, Airflight Ltd. Ever versatile he later founded Fairthorpe Ltd., a company supplying sports cars in kit form, which he owned until 1983.

Donald Bennett died 15 September 1986, survived by his wife, son and daughter, remembered as a superb aviator whose subsequent career could be termed as a disappointment, bedevilled as it was by relentless government red tape and mindless regulations in his post-war endeavours.

**Lancaster III (103 Sqdrn.)**

# Australia's Flying Scotsman

**G.U.(Scotty) Allan**

# George Urquhart (Scotty) Allan 1900-1996

In a career devoted to aviation George Urquhart (Scotty) Allan emerges as a most remarkable airman. His aerial saga began as a 17 years old tyro thrust into combat on the Western Front, followed by a hectic period as a pilot in the post-war RAF and from there to a decade of involvement in Australian civil aviation, until the outbreak of World War Two, which saw further service, this time as a wing commander in the RAAF.

Scotty Allan remains less well-known than his contemporaries, names redolent with adventure and courage, typified by Charles Kingsford Smith, Charles Ulm, Bert Hinkler and a host of others. Scotty flew aeroplanes extremely well, he understood the business end and the engines better than most, and while others were making capital and taking in the aura of heroes, Scotty got on with the job, showing a preference to remain in the background.

Born 2 February 1900, the third youngest in a family of six brothers and two sisters, Scotty's early years were spent in a countryside environment in Perthshire, Scotland. At age seven his interest in aviation was kindled by a local man's efforts to build and fly his own aeroplane. The man was no doubt inspired by the Wright Brother's successes, however his machine never did fly, but it did make a lasting impression on young George, excited by the possibility that a heavier-than-air machine could actually fly.

Scotty was still a schoolboy at the outbreak of war in 1914 and to avoid conscription; in 1917 he travelled to Glasgow to enlist. Quite by chance the recruiting office he applied to was accepting entrants to join the Royal Flying Corps. He passed the entrance examination with ease and after ground instruction he became a flight cadet at Eastbourne, where he survived a rigorous flight training period on the quirky Sopwith Camel. Still only aged seventeen Scotty was posted to a Camel squadron in France, flying a number of operational sorties. He also flew the DH4 on photographic and bombing missions, and after one particular sortie he was found to be under age for operational service. From then until the war ended in November 1918 Allan became a ferry pilot, delivering aircraft between England and France.

In a post-war situation Scotty found employment as an engineer with the Ford Motor company, later moving on to Rolls Royce for a period of two years. In 1923 he re-enlisted in the RAF as a ground engineer, but without revealing his tenure as a pilot during World War 1. In due course he was promoted and posted to Iraq, where the RAF was entrusted with maintaining the peace.

Eventually Allan was accepted for pilot training, where understandably he did very well, and on gaining his wings as a sergeant pilot was posted to Helwan in Egypt, operating on the DH 9A. After further service at Khartoum, his Middle East tenure ended there, which saw him posted back to England. He served in a variety of squadrons, one of which was No.58, a heavy night-bombing unit, operating the lumbering Vickers Virginia.

Around this time Allan and a squadron companion responded to an advertisement in the weekly 'Aeroplane' journal where two pilots were required for an Australian airline: 'Used to heavy bombers and able to fly at night'. The ad was lodged by Charles Ulm, who with Charles Kingsford Smith was assembling a team of pilots and engineers to man their emerging airline, 'Australian National Airlines'.

Based in Sydney their operation began 1 January 1930, with flights from Sydney to Brisbane, Sydney to Melbourne and later Melbourne to Launceston, using a fleet of tri-motor Avro Xs. The company soon established an enviable reputation for safety and reliability; however on 21 March 1931 tragedy struck with the loss of 'Southern Cloud' on a scheduled flight from Sydney to Melbourne. ANA suffered a body blow to its reputation with the resultant adverse publicity and costs in the fruitless search for the missing airliner and later in the year closed down its operation.

As a result Scotty was made redundant, but over the ensuing three years he acted as pilot for Charles Ulm, who had purchased 'Southern Sun', renaming it 'Faith in Australia'. They made many flights; barnstorming in New Zealand plus an epic flight to England via India and the Middle East. Their partnership ended in December 1934, when Ulm's Airspeed Envoy vanished between California and Hawaii on the first stage of a global survey flight.

Once again Scotty faced unemployment; however a far more secure opportunity came his way, by virtue of an invitation in 1934 from its creator, Hudson Fysh to join the burgeoning Qantas airline as a senior pilot. Qantas had recently entered into an agreement with Imperial Airways whereby Qantas would fly outgoing passengers and mail to Singapore and from there Imperial would deliver those passengers and mail to England and vice versa, with Qantas taking the incoming passengers to Australia.

This service was based around the employment of the new DH86 airliner, which came under scrutiny following the loss of Qantas VH-USG on its delivery flight over central Australia. Two other DH86s, operated by Holyman's Airways were also lost, in this instance over Bass Strait. For the time being the Qantas

DH86 fleet was relegated to air mail duties until the type was eventually certified for passenger service.

Qantas continued to operate the DH86 in conjunction with Imperial until 1937, when they were gradually replaced by the iconic 'Empire' flying boats. These graceful craft offered new standards of comfort on Imperial routes to Australia, India and South Africa; however the outbreak of war in September 1939 had the effect of revising these routes as hostilities became more global.

In 1940 Scotty Allan enlisted in the RAAF, and after holding the rank of pilot officer in various squadron detachments, he was despatched to The United States on a mission to acquire suitable combat aircraft for Australia's defence. America at that stage was not officially at war but was still prepared to offer aircraft and materiel to Britain and her empire on a lend/lease arrangement. Scotty's mission had a positive outcome, with an agreement for America to provide a regular flow of PBY Catalina flying boats for RAAF service. In fact the first example, captained by Allan reached Australia in February 1941.

That same month he was posted to the Seaplane Training Flight at Rathmines on Lake Macquarie south of Newcastle to set up a Catalina training establishment. In this he was most qualified, rising through the ranks to wing commander. In December 1944 Scotty was discharged from the RAAF and rejoined Qantas. Although the European war was yet unfinished, Qantas and BOAC were already making plans for a post-war return to civil aviation.

Scotty was despatched to England to resume contact with various manufacturers, in particular Avro, who had created the Tudor airliner, which hopefully would be accepted by Qantas and BOAC. Scotty tested an early example, and in a report to Hudson Fysh, Qantas general manager, he declared that the Tudor would only incur a loss in operational use, and that a better option was the Lockheed 749 Constellation.

As a result Qantas cancelled their Tudor order, leaving Scotty to relay this unpalatable news to the manufacturers, Avro and to BOAC, who had already placed a substantial order. Avro's attitude was quite reasonable, not so BOAC, who accused Scotty of incompetence in his decision. As it happened, the Tudor never entered BOAC service, following a series of unexplained losses with other operators, although Scotty never did receive an apology from BOAC.

Meantime, Qantas as with other operators soldiered on with re-furbished ex-US Army DC4s, until 1948, when with Commonwealth Government approval their fleet was bolstered by four Constellations. However a new element had entered the civil aviation scene; this was the first jet airliner, the DH 108 Comet.

This ground-breaking aircraft first entered commercial use in May 1952, with services from the UK to Cape Town. BOAC was the initial operator, and following optimistic reports on its performance, plus pressure from the British Government Qantas despatched Scotty to England to make an assessment.

He flight tested an example, accompanied by De Havilland's chief test pilot, the renowned 'Cat's Eyes' John Cunningham and expressed unease about how sensitive were the controls; as though the speed would build up and the machine could likely break up in the air. Another factor, of which Allan was critical, was the outer plating. The edges were all arranged in a long line, rather than staggered to prevent the continuation of any tear that might occur, so that the fuselage doesn't rip completely open. Scotty relayed his findings to Hudson Fysh, and with a recommendation that Qantas should not buy the Comet. Fysh trusted Scotty's judgement, and despite pressure from the British Government, cancelled the Comet order, and instead ordered eight examples of the Lockheed 1049 Super Constellation, which served Qantas exceedingly well.

Scotty's suspicions of the Comet were well vindicated with the loss of three Comet1s; the first out of Calcutta in late 1953, followed by two more over the Mediterranean Sea during 1954. One of these, G-ALYP was recovered from the ocean depths and the scattered wreckage brought to the surface. The cause of the failure was accurately pin-pointed as structural failure at the corner of the rear RDF window.

Qantas still had to face hard questions from the Commonwealth Government as to their decision not to order the Comet. These were overcome and as a consequence, permission was given to go ahead and order the Boeing 707, with the first delivery happening in Sydney in 1958. Since that memorable event, Qantas and Boeing have forged an enduring relationship, with their on-going use of classic Boeing types.

Scotty's employment with Qantas ended with his statutory retirement in 1961, after holding key managerial positions. He suffered a heart attack in 1972, which entailed immediate surgery. Following a slow recovery, Scotty still led an active retirement until his death in Sydney in 1996.

In 1979 Qantas pilot Lester Brain made a faithful summation of Scotty Allan in a personal letter, which concluded thus; *You toiled and wracked your brain and staked your reputation in creating Qantas into the great international airline of which all Australians are so proud.*

**VH-ABG 'Coriolanus' 1938**

## Author

Born in Sydney Australia Murray McLeod is an accredited artist specializing in landscape, aviation and portrait subjects. He has been a regular contributor to an international on-line magazine with his aviation and motorcycle articles. Murray was also an original contributor to South African based Global Aviator on a monthly basis with his illustrations and articles

He has illustrated books published on his favourite themes, vintage aviation and historic motorcycle racing. Murray welcomes commissions from interested readers for that special art work.

Author website

www.mcleodart.com.au

## An appreciation

Sadly, none of those trailblazers are with us today. Scotty Allan's passing in 1996 represented the last of that host of heroes, although none would ever lay claim to that title. In war and peace they gave of their best, often in circumstances that seemed insurmountable and in machinery whose qualities were generally inferior to the determination of their crews.

www.ingramcontent.com/pod-product-compliance
Lightning Source LLC
Chambersburg PA
CBHW051152220526
45473CB00003B/744